Project Studios: A More Professiona

I dedicate this book to
Marta Izquierdo Rojo

Project Studios: A More Professional Approach

Philip R. Newell

Focal Press

OXFORD AUCKLAND BOSTON JOHANNESBURG MELBOURNE NEW DELHI

Focal Press
An imprint of Butterworth-Heinemann
Linacre House, Jordan Hill, Oxford OX2 8DP
225 Wildwood Avenue, Woburn, MA 01801-2041
A division of Reed Educational and Professional Publishing Ltd

 A member of the Reed Elsevier plc group

First published 2000

British Library Cataloguing in Publication Data
A catalogue record for this book is available from the British Library

Library of Congress Cataloguing in Publication Data
A catalogue record for this book is available from the Library of Congress

ISBN 0 240 51573 0

Typeset by Avocet Typeset, Brill, Aylesbury, Bucks
Printed and bound in Great Britain by Biddles Ltd, Guildford and King's Lynn

Contents

About the author

Philip Newell began work in the music industry in 1966 as a live sound engineer in his home town of Blackburn, England. In 1968 he moved to London, where he worked with many well-known artistes and in 1969 built Majestic Recording Studios. In 1970 he moved to Pye Recording Studios in Central London, which was then one of the country's major studios, and in 1971 joined Virgin Records as chief engineer. He became technical director of Virgin Records in 1973 and held the post until he sold his shares in Virgin in 1982 to concentrate (briefly) on his seaplane operations. In 1984 Philip founded the recording division of Reflexion Arts with Alex Weeks, but left in 1988 to work as an independent acoustics and electro-acoustics consultant.

At the time of the publication of this book Philip Newell had designed over 150 recording studios, together with many cinema dubbing theatres, music clubs, discotheques, concert halls, television studios and mobile recording vehicles. Throughout the past 30 years he has been involved in innumerable recordings as systems engineer, recording engineer and producer. His experience of the recording processes covers the artistic as well as the technical sides and he has recorded an extremely wide range of music, from solo instruments to big bands, orchestras, rock bands and many other forms of music in many different countries.

Philip currently lives in Spain, which has been his home since just before the European borders came down in 1992. He is a member of the Institute of Acoustics, a member of the Audio Engineering Society, and since 1987 has been closely involved in much work at the Institute of Sound and Vibration Research (ISVR) at Southampton University, England. At the ISVR he has sponsored numerous students during undergraduate projects, post-graduate work and doctoral research on the subjects of studio monitoring and sound absorption systems.

The author produced Mike Oldfield's famous live double-album 'Exposed' and in 1976 remixed 'Tubular Bells' for a quadraphonic release. In parallel with being responsible for the construction of studios such as The Townhouse in 1978 he was also closely involved with the design of numerous private studios in musicians' homes and made many 'on-site' recordings with mobile recording studios. With The Manor Mobile in 1973 he made what is believed to have been the world's first live recording by a purpose-built 24-track mobile recording vehicle.

Philip Newell is the author of over a dozen papers presented to the Institute of Acoustics and the Audio Engineering Society and has written around 100 articles for the music recording and aviation press. He was formerly also an examiner of seaplane pilots and flew in many air displays. One day he intends to get himself a steady job.

Credits

Chapter 3 includes text taken from the following article:
Macatee, Stephen R., 'Considerations in Grounding and Shielding Audio Devices', *Journal of the Audio Engineering Society*, Vol. 43, No. 6, pp. 472–83 (June 1995).

Chapter 9 includes text taken from:
Giddings, P., *Audio System Design and Installation*, Focal Press, Boston, USA and Oxford, UK (1995)

Figure 4.2(b) is taken from Lewis, Warren H., 'Poor quality and grounding', *Sound and Video Contractor*, pp. 16–50 and p. 73 (September 1997).

Figure 4.4 is taken from Glassband, Martin, 'The Origin of Balanced Power', *Sound and Video Contractor*, pp. 54–60 (September 1997).

Figure 7.4 is taken from Harrison, Peter M., *Sound Pro*, Vol. 1, No. 5, p. 29 (February 1998)

Acknowledgements

I would like to thank Janet Payne for the very great efforts that she has made throughout the writing of this book. She has been fundamental in the production of the three books that I have written to date. I could not have done them without her.

Thank you, again, to Sergio Castro AMIOA for his tireless work on the figures during all their reiterations. I really appreciate the effort.

Thanks also to Dr Keith Holland, for the generation of many of the plots and for the hours of discussion which go into the preparation of a book such as this.

Somewhat belatedly, I would like to recognise the patience and welcome advice of Keith Spencer-Allen. Keith, on behalf of Focal Press, has acted as 'reader' of the drafts of *Studio Monitoring Design*, *Recording Spaces* and now *Project Studios*. Looking through my files I see that he has made about 150 comments and suggestions, on only three of which I have failed to act. Keith's contributions have been very instrumental in shaping the direction and improving the quality of the three books. I value greatly his contributions.

Philip Newell
Moaña, Spain
July 1999

Preface and synopsis

The 'call' to write this book came from the finding, from my extensive travelling, that so much of the international recording industry is now operating in a world filled with misconceptions and marketing hyperbole. This is doing nothing to sustain the professional approach to recording that has been built up by so much hard work over so many years, by so many people. In the world of aviation, empowerment to buy an aeroplane does not imply empowerment to fly it. Lives would be at stake, so society places restriction on flying, and imposes minimum qualifications on pilots. In the music industry, though, where lives are *not* at stake, empowerment to use is usually seen as an automatic result of empowerment to buy. Small studios often expand, and dubious attitudes and philosophies can grow with them. One result is that I have now seen many, rather large studios in the hands of personnel who were pitifully under-experienced when it came to obtaining the full potential from their rooms and equipment; and subsequently from the musicians.

Project Studios is not a book about recording techniques, or how to use various items of studio equipment. It is a book which deals with the fundamental realities of an industry, its psychology, and the electrical and acoustics laws which it cannot force to compromise. The intention of this book is to take the reader through the evolutionary processes which have brought us to where we are, and to understand the driving forces behind that evolution. The book also deals with the fundamental technological requirements to get the best performance out of recording equipment in general, and also looks at some areas of understanding which are often found to be lacking. It has been said that anything which seems obvious in acoustics is probably wrong. A considerable proportion of the book is thus concerned with the signposting of many such misleading pathways, and it gives sufficient background information and further references to help the project studio operators to avoid being misled by hearsay or erroneously perceived truths.

Project studios now form a major proportion of a huge world-wide industry, but very often they exist in a twilight world of understanding. The information contained in the book should help to illuminate many of the less accessible areas of knowledge, much of which can be very difficult to find in a concise and readable manner. The information may also be of great use to many people in the fully professional studio world who would like ready access to some aspects of information which will benefit them in their daily work.

The book begins with an introduction which generally sets out the purpose the book seeks to fulfil, and explains why its contents are so relevant to this expanding face of recording.

Chapter 1 – The evolution of the project studio

The first chapter takes us through the story of many producers and musicians seeking to wrest more control over the music recording process during a time when the operational requirements of the limited equipment necessitated the attendance of technical personnel, who were not always as co-operative or as understanding as they could have been. However, as the technology advanced, groups of 'pioneer' musicians set out alone to exploit their new found freedoms, sometimes only to find out that they had bitten off more than they could chew. Ultimately, it is largely this pathway that has led us to our current situation, where the 'brains' of the recording process are often locked inside the equipment. This may, or may not, be seen as a step forward, dependent upon circumstances.

Chapter 2 – The equipment and the compromises

Obviously, if one only has a budget for a Ford Fiesta, then one cannot reasonably expect to have a Rolls-Royce. In the world at large, such a statement would seem to be common sense. However, in the world of the project studios, such logical reasoning does not always prevail, and equipment marketing exercises often tend to encourage the belief in unrealistic dreams. Chapter 2 looks at what can be achieved for 'realistic' prices, and where restrictions on performance must be expected, even if they are quite deeply hidden. By knowing the limits, one is better positioned to extract the maximum achievable performance from any given device. By knowing what is reasonable to expect, dreams are not so easily built up – or shattered.

Chapter 3 – Interfacing

With so many manufacturers chasing the huge project studio equipment market, many corners are cut, and many systems are re-invented in order to gain some perceived market edge. These developments are often made in circumstances of great commercial secrecy, and equipment is frequently produced by different manufacturers with a growing number of different concepts of input and output configuration. When much of this equipment is inserted into chains of top line recording equipment, as is often done to demonstrate the new equipment's capabilities, all is usually well, due to the very tolerant, *and* expensive interfacing ports of the professional chain. Unfortunately, however, when realistic project studio systems are assembled from such diverse mixtures of equipment, having idiosyncratic interfacing topologies, chaos often awaits. What may be worse, poor interfacing may not be obvious for what it is, and may lie hidden, limiting the achievable quality of a whole chain without the operators ever being aware that by a simple change in wiring regimes much greater sonic quality could be realised.

Chapter 4 – Mains supplies and earthing systems

Somewhat similar to the problem of inappropriate audio interfacing, the precise form of power supply and earth wiring can cause sonic degradation. However, they can also cause computer crashes, system glitches, noise problems and control system irregularities. Unlike the audio interfacing, though, they are usually controlled by the laws of countries or regions, and these human laws, whilst being *legally* inviolate, do not always go hand-in-hand with the requirements of the physical laws which govern the application of electricity and earthing to current studio systems. Nevertheless, for optimised performance, the two sets of laws *must* be respected, though their violation is widespread. This chapter attempts to set out the problems and their solution in an easily understandable fashion, and makes clear statements to engender an awareness of the causes and effects of the erroneous wiring systems which are perhaps more likely to be found in project studios than in professionally installed systems.

Chapter 5 – Monitoring requirements

Of course, in order to be aware of many problems of sonic degradation, and especially in order to be able to hear when subtle improvements have been made, one needs a certain minimum degree of resolution in the monitor chain. In so many project studios, these conditions simply do not exist, but it is often due to pure ignorance that higher resolution systems are not employed. Once the reasons for high resolution monitoring have been established, it follows that its use as a diagnostic tool to assess the rest of the signal chain will become more apparent. Good monitoring, in itself, can be a great teacher. This chapter points out many of the frequent misunderstandings and misconceptions which apply to many common project studio monitoring systems and environments. It shows how good systems can not only lead to better recordings, but can also save money by allowing existing equipment to be used optimally, reaching standards of performance perhaps thought to be only achievable by investing in more advanced devices.

Chapter 6 – Modular digital multitrack recorders

The spread of the ADAT and 'DA88' type digital tape recorders has revolutionised the whole world of project studios. As with much equipment which is principally designed for use in project studios, these machines can reach excellent heights of performance, but are nonetheless built to a price, and that fact implies that their overall performance will not match the fully professional machines costing 400 per cent more. Chapter 6 is based on five years of personal experience with such machines, gleaned from their use in a very wide range of circumstances. It also deals with operational and philosophical aspects of their use and maintenance, together with some rather disgraceful faces of the manufacturers which are only revealed when their users are in trouble. The chapter is illustrated with photographs of the machines in use in some extraordinary situations. It also gives some very practical advice, and looks to their probable future.

Chapter 7 – What's going on? Recording levels and metering

In all too many cases, and quite reasonably from the point of view of what perhaps should be expected, project studio users are often grossly misled by believing too literally what the various metering devices on their equipment are telling them. Over the years, signal level meters have been developed for some very specific purposes, and we now find ourselves in a situation where an unreasonably wide range of metering standards are in use within many single signal chains. The chapter investigates some of the thinking behind the various standards, the actual levels that they are designed to emphasise, and the things which need to be borne in mind when connecting together equipment using different metering concepts. The relevance of the metered levels to the actual recorded levels is also investigated.

Chapter 8 – Mastering

I know a lot of top mastering engineers, and they all, privately, admit that they should be running out of work. In reality though, they now seem to be needed more than ever, and are being elevated from 'back-room-boys' to being the high-flying saviours of our industry. Chapter 8 explains what the mastering profession is, why it is difficult for project studio users to do their own mastering, and why the special talents of the mastering engineers are so necessary. The chapter includes interviews with very experienced recording engineers/producers and mastering engineers, giving a rare insight into their specialised world. Their experiences of what quality to expect from what sources, and when it is, or is not, possible to do anything to improve results, makes essential reading for many people who cannot currently gain access to top line facilities, yet who, somewhat ironically, are the ones who most need them.

Chapter 9 – Some further items for consideration

This is a chapter about some little discussed aspects of recording which, nevertheless, have great bearings on recorded quality and the smoothness of operations of the whole process. The essence of the chapter is based on my April 1988 article in *Sound Engineer and Producer* magazine entitled 'The Duff Leading the Deaf'. As that situation still seems to be very widespread, it seems worth bringing the piece up to date, and spelling out the folly inherent in much erroneous thinking.

Chapter 10 – Some basic acoustics of loudspeakers and rooms

Of course, for one to be able to make appropriate judgements of recorded quality, one needs a monitoring system. Chapter 5 discusses some of the basic points, but the level of comment on loudspeaker performance that one hears in many project studios is naïve in the extreme. Loudspeakers in rooms, as an ultimate means of assessment, are very blunt tools, but they are more or less all that we have. What is discussed in this chapter is essential before we can move on to the following chapters, which discuss the principles behind stereo and surround, and describe how rooms and loudspeakers can best be configured to make working with these formats as enjoyable and productive as possible, without necessarily breaking the bank.

Chapter 11 – The small room problem

Following on from the ideas of the previous chapter, it describes how small rooms, of the size typically found in project studios, can be made to perform like some of the best rooms around, without spending fortunes or requiring specialised skills to construct them. The techniques discussed have been employed in many countries over many years, and are well proven.

Chapter 12 – Stereo, the unstable illusion

When we have completed the installation of good loudspeakers in good rooms, then all should be sweetness and light, shouldn't it? Well, realities are a different thing, and this chapter looks at some of the aspects of stereo itself which are frequently little understood, yet which cause frustration daily by making their disturbances obvious whilst leaving the origins unknown to the recording personnel. Many people seem to think of stereo as a sort of birthright, yet it is nothing but a very unstable illusion.

Chapter 13 – Phase, time and equalisation

Why do equalisers sound different when set to the same responses? Why can room problems not be equalised by conventional equalisation? Why were such systems erroneously employed in the past? What is the future for digital control of such problems? Many of the answers lie in the preservation of waveform responses, which if not respected, will conspire to confound many attempts to make good recordings. It is the phase and time responses inherent in any recording system which hold the key to so much naturalness in sound quality, but whose relevance is all too often ignored. Chapter 13 explores the worlds of phase and time, and tries to answer many of the questions which they raise.

Chapter 14 – Computers in control rooms

Almost all sound recording control rooms now contain computers of one sort or another. In fact, computers now form the hub of many recording systems, but their dominance has caused their presence to become obtrusive in many situations, where their introduction into the recording systems has been a rather rude affair. This chapter looks at many of the things which computers have to offer; a good number of which are simply unavailable in the analogue domain. However, it also discusses how to keep the computers subservient to the whole recording process, and not vice versa. Computers are here to stay, but it is better for everyone if they behave like good neighbours, and not noisy ones.

Chapter 15 – Considerations on music-only surround: can we control the chaos?

Is surround the way of the future? Only time will tell. Budget priced systems are already available for project studio use, but if the previous chapters have shown stereo to be anything *but* simple, then its problems can sometimes be small compared to surround. The chapter looks at how to prevent a repetition of the debacle of quadrophonics in the 1970s, and shows clearly which recording practices will lead to heightened enjoyment, and which others may lead to unmitigated disaster.

Chapter 16 – Horns: their strengths and weaknesses

Much maligned and widely misunderstood, horn loudspeakers play a vital role in many widely used loudspeaker systems. A look at some of the facts about their properties and application allows us to discuss a number of aspects which relate to monitoring and sound reproduction in general. When horns may be the best solution to a problem, prejudices should not discount their eligibility, or deprive us of their benefits.

Chapter 17 – Foldback

Whilst the previous seven chapters have looked carefully at the world of the control room, it must be remembered that when musicians are performing, their own world may be a very small one, constrained within a pair of head-phones. If this world is under the control of technical staff in a control room, then they may not be aware of all the needs of the musicians. If these needs are not fulfilled, then the poorer will be the musicians' world. Under such circumstances, it will be unlikely to get an optimum performance, and consequently, the recording will suffer. Chapter 17 looks at different concepts of foldback systems, their ranges of use, their strengths and weaknesses, and their appropriateness to different circumstances.

Glossary

The book concludes with a glossary of a selection of the terms used, and brief descriptions of their relevance.

Introduction

What have become known as 'project studios' have developed in a somewhat grey area between the most rudimentary of home studios and the fully professional studios. The dividing lines between amateur and professional operations have blurred to such an extent in recent years that a continuum now exists from the bedroom or garage studios to the world's finest facilities. With the bridging of the old divide has come a mobility of personnel, crossing from the amateur to the professional operations, whose apparent capabilities are not always accompanied by a solid understanding of their working principles. This situation can have its beneficial effects, such as bringing new ideas into the industry, but the benefits are often outweighed by the drawbacks. Misunderstanding the basic principles can lead to bad recording practices, poor decision making, misapplication of equipment, and the inability to get the best performances out of the musicians.

The above limitations can lead to the failure to realise the full potential of many recordings, not only in the technical sense, but also in the artistic and performing senses. All too often, if nobody is available to demonstrate time honoured practices, many people involved in project studio recording never realise what *could* have been achieved from their endeavours. When recording staff are trained in the larger, commercial studios, people are on hand to offer advice when needed, but so many of the project studios operate with only minimal staff. Frequently, also, experienced advisers cannot be afforded by such studios, so, despite the project studios occupying a significant place in the current recording industry, they can be a rather unpredictable area of operation.

It was in the 1970s when the first, true, home recording equipment began to appear. What was available in those days was no match for the then current professional equipment, but was welcomed by many musicians. It allowed them to write music and to prepare ideas in a more relaxed atmosphere, and also allowed them to make demonstration recordings without the 'clock watching' caused by the financial pressures of having to use commercial studios. The home equipment also allowed the musicians to work when they felt inspired to do so, and not at a perhaps inconvenient time, determined by the availability of a studio.

The market for home recording equipment grew to such a size that it caught the interest of many manufacturing companies. Realising the size of the market, and hence the money which it could generate, the companies producing the equipment became involved in a race to develop the technology, with huge rewards beckoning to those who could lead on the cost/performance front. The improved and readily affordable equipment which began to result from this race caused the market to grow even further, promising the manufacturers even more money, so the development race accelerated. In a remarkably short space of time, between the mid 1970s and early 1980s, the performance gap between amateur and professional equipment had diminished to quite a considerable degree.

During this same period of time, a great number of recordings were being made by the mobile recording facilities which were then available. Many musicians had opted for an out of studio recording location, where they felt more creative and relaxed. In many cases, after successful demonstration recordings had been made on 'home' equipment, professional mobile recording facilities were brought to the same locations in which the 'demo's' had been made to record an album. This new trend broke the mould of the traditional ways of doing things, and by the mid 1980s, had opened the flood gates to a whole new approach to recording.

Certainly in the countries with highly developed recording industries, moderately successful artistes could reasonably expect $50,000 to $100,000 (US) as a recording advance for an album, which had traditionally been for the cost of recording in a suitable professional studio. However, the rapidly advancing quality of the then current 'semi-professional' equipment, together with its falling price in real terms, had reached a stage where a group of musicians could build their own quite respectable studio from the recording advance. By this means, once the album was finished, they owned their own studio, which was a dream come true to many of them, but dreams are one thing, realities can be something else. What is more, in dreams, things are not always as they seem.

The recording advances, when used in a front line studio, bought not only the use of some very good recording equipment, but also the ability to rely on a highly trained staff. Recording equipment was often checked daily for its optimal operation, and any malfunctions could usually be dealt with rapidly by the technical staff. Sound isolation, not only to the outside world, but also between the studio and control room, ensured the minimum of disturbances and the cleanliness of the recordings. These professional studios were expensive to hire, but good sound isolation, good internal acoustics, and their highly trained staff did not come cheaply. What is more, in a professional studio, machinery has to work, hour after hour, day after day, and year after year. This degree of reliability is not cheap, either.

Whilst it is true that many excellent recordings were made outside of the studios by the mobile recording facilities of many recording companies, and that these were often done in unprepared locations, it should not be forgotten that they were professional facilities. They had professional equipment and, perhaps more importantly, professional and highly specialised staff. The staff could look out for problems in the recording environment, and were well versed in recording in unprepared surroundings. Their very professionalism was what made it all look so easy, but, often unbeknown to the musicians who were being recorded, the well-known studio engineers who may have been engineering the sessions frequently relied heavily on the advice of the specialist mobile crews with whom they were working.

Unfortunately, so much of the 'behind the scenes' functioning, and the underlying reasons for things being the way that they were, went unseen by many musicians. The temptation of the prospect of a 'free' personal studio when the album was completed drew them to the conclusion that the equipment was 90 per cent of the recording story, so they opted to spend their advances on the newly available equipment, and to dispense with the traditional route to recording an album. The repercussions of this new trend were enormous, and have subsequently changed the face of the whole international recording industry. The three most significant effects were, firstly, to take away so much of the available work from the established studios; secondly, to cause great cut-backs to be made in the staffing levels of existing studios; and, thirdly, to put enormous power into the hands of equipment manufacturers, who now lead much of the industry by the nose.

The truly professionally operated studios have become something of an elite echelon, and there are many countries, even in Europe, where they do not exist at all. Within the top-flight studios exists a huge amount of knowledge and experience, where things are perhaps better than they have ever been, but they now produce only a tiny fraction of the commercially produced recordings, as opposed to the great bulk of the commercial recordings which they produced twenty years ago. If only some of this knowledge and experience could easily filter down to the less developed section of the industry, which now forms its core, it could be greatly productive, but, as time passes, there seem to be fewer and fewer channels through which this information can flow.

As the face of the recording industry has changed, so has the form of the press and publications which are associated with it. Magazines used to exist which were written *by* professionals *for* professionals. They were forums for open discussion between professional peers. For people aspiring to join the industry, or advance their experience, these journals were required reading, and people used to look forward to the monthly publication dates. The publications were usually in the hands of small groups of highly dedicated editorial staff, and as they were read each month by a good proportion of the industry's top professionals, equipment manufacturers and distributors sought to advertise in these publications. The advertising was important; it provided much needed revenue, as the cover price of the monthly issues could hardly support the publishing costs on their own. The important point is, though, that the text and the advertising were independent.

The changing face of the balance of commercial studios has, in turn, led to a very differently targeted recording press. A whole new 'home recording' market has grown at an incredible rate, spawning its own wide range of magazines. In general, these publications are of great use to the home recordist, but most of the articles which they contain tend to be somewhat short and superficial. They obviously take into account the fact that they are not facing a highly experienced readership, though many of their readers may well be very skilled in certain aspects of the recording art. At the other extreme we have publications such as the *Journal of the Engineering Society*, but this is way over the heads of even many of the best recording engineers, and has, of late, become much more academic in its content. Without a good grounding of university level mathematics, it is almost impenetrable. The industry is still well covered by a range of magazines, but they do not necessarily overlap. Most of the publications are now in the hands of companies who are not dedicated to publishing, but to profit. The upper level of the studio league table is now too small a part of the whole industry to warrant its own, dedicated magazines. Advertisers want to sell their products in quantity, so they usually only opt for the services of magazines which reach the widest range of potential customers. This trend has tended towards multi-media magazines, and a heavy bias towards articles which relate to new equipment, which seems to appear on the market at an ever-increasing rate.

Technology is a popular subject for discussion, and magazines both like and need to deal with popular topics, but technology is not the be-all and end-all of recording. It is technology, however, that can be mass produced, and which sells in large quantities. It can therefore give huge financial leverage to the equipment producers, and there is a whole section of the recording studio industry which is now almost enslaved by the technology. This is a wonderful state of affairs for the equipment manufacturers, for many of whom the relative ignorance of a considerable proportion of their customers, at least on the wider aspects of recording, is a great bonus, as it allows the endless promotion of updates and accessories, which may or may not be as necessary as they seem.

So many things in modern life are becoming rather more transient than they used to be, and boxes of equipment fit in well with this state of affairs. Boxes can be bought, sold, exchanged and up-dated, as standards and fashions change. Building special rooms for recording, on the other hand, and acoustically treating control rooms, tend not to be seen to be such negotiable investments. Strangely, also, in a world which is so dominated by marketing, good rooms are often not seen to be so 'attractive' to a whole section of studio operators and clients, who live and work in an almost entirely equipment and software based world. As so much of their education in this area comes from hearsay and advertisement dominated magazines, the promotional hype and myths tend to become self-perpetuating. This all may sound a little negative, but it is intended to be an observation of the way in which an industry has become as much driven by its suppliers as by its market or its own inherent evolutionary forces.

However, the mid 1990s saw a swing back to the recording of 'real' instruments, and much of the latest semi-professional equipment is capable of truly excellent sound quality. With all this in mind, the time is perhaps now right to dedicate a book, not to general recording practices for the home recordists or small studio operators, but to getting the very best from the potential which the current project studios *can* achieve, especially in the right hands and with the right artistes. What is contained, herein, is much hard earned experience, and much behind-the-scenes insight. First, however, some scene-setting will be needed, so we shall need to look at how this whole business has split off from the mainstream, which is the subject of Chapter 1.

Chapter 1

The evolution of the project studio

The evolution of music recording has taken a path which I would never have expected when I first became involved in this industry in the mid 1960s. Rather than developing into the huge professional industry that I had imagined, a whole middle order of the industry has become dominated by the so-called 'project studio'. Looking back, though, the origins of the development of the 'project studios' were already in place long before anybody recognised them as having any relevance whatsoever to main-stream recording. In the immediate post Second World War period, Bing Crosby was one of the driving forces behind the development of magnetic tape recording for music, because he, as a musician, wanted more control over the process of his own recordings. In the mid 1950s Les Paul and Mary Ford pioneered multitrack audio recording, by asking Ampex to produce an 8-track tape recorder with off-the-record-head monitoring for track synchronisation. Around the same time, in the UK, Joe Meek was straining professional sensibilities with his own pioneering approaches to some very non-standard recording techniques.

Such ventures outside of the realms of the somewhat clinical, professional recording practices of the day were, at that time, only for the rich, the famous, or the brave. However, by the early 1970s, bands such as the Rolling Stones and Led Zeppelin began taking mobile recording facilities to out-of-the-way locations which they found to be more conducive to achieving an optimal performance of their music, but still they used what was state-of-the-art equipment, *and* well-trained recording engineers.

Until that time there existed two well-defined strata of recording studios: the professional studios, and the 'demo' studios. The latter were almost exclusively used for making demonstration recordings to take to record companies for the purpose of gaining their interest. In almost all cases, if the record company was impressed, the recordings would be re-made in professional studios, and only in very rare cases did the recordings from the demo studios reach the shops.

It was only around 1970, when tapes such as Scotch 202 and Ampex 434 became available, that the recording process began to be freed from its strait-jacket of precise level controls. The window of acceptable recording levels had been not much more than a crack: too much level brought the onset of rather nasty distortion, and too little level lost the music in the tape noise. Professional recording was *not* a game for amateurs, and the quality of some of the pre-1970 recordings is a testament to just how good many of the recording engineers were. They were faced not only with the restriction of tape which was 'unusable' by the standards of today, but also with daily equipment failures and alignment drifts. It was not uncommon for the number of maintenance staff to equal the number of recording staff. 'One-person' operations were virtually out of the question unless the equipment was very simple; in which case it would probably only be suitable for the most basic of recording methods. Most serious recording necessitated an adequate staff of trained personnel, so almost automatically required the services of a professional studio.

1.1 The standards of the day

My own experiences of the off-shoot of a serious, musician-only oriented recording industry began in 1971, when I was being lured to The Manor Studio, near Oxford, UK, by its then managing director, Tom Newman. The story serves well to highlight the development of parallel tracks. Tom had some 4-track BASF tapes recorded on a B&O domestic tape recorder, which were demo's by a relatively unknown artiste, Mike Oldfield. Tom had boundless faith in the prospects for these recordings, and promised Mike that once Virgin Records had a usable recording studio facility, he would be one of the first people to be recorded for their proposed record label.

The Manor opened for business on 3 November 1971, and about three weeks earlier I had joined the company as chief engineer; moving from a much better paid job at Pye Recording Studios, in London. I liked the concept of The Manor, though, and the prospect of living in a country manor house was definitely attractive after spending years travelling through the London rush-hours, day after day. Most of The Manor's equipment was 'marginal' in terms of the then current professional standards, though its Ampex MM1000, 2 inch, 16-track tape machine was without equal at the time. One of my first tasks was to try to squeeze the last drop of performance out of some of the old secondhand or marginally professional equipment, but at times I felt like I was swimming up a waterfall.

Tom Newman was the managing director of the company, but had no formal recording training; he was essentially a musician. As quickly as I was trying to up-grade and 'professionalise' the proceedings, Tom was introducing new 'embarrassments'. Looking back, some of my embarrassment was, no doubt, down to professional snobbery and to some insecurity, but we were trying to launch an ostensibly professional operation on a rather shaky foundation, and we needed the confidence of the record companies or we would not get the work that we needed to survive. Whereas *I* was trying to keep things up to accepted standards which the record companies would recognise as being suitable for commercial recording, Tom was still thinking of the studio as being primarily for musicians, and *their* needs.

To me, with my formal training, this seemed to be an unprofessional and unviable approach to the concept of recording. Tom, on the other hand, saw the professional studio approach as being insensitive to the needs of the musicians. We were, and still are, the best of friends, but we had come from very different backgrounds. What he saw as being insensitive to the needs of the musicians, *I* saw as being a realistic approach to the technical aspects that I thought he was viewing over-simplistically. *He* wanted a studio for musicians, which is ultimately what The Manor became famous for being, but in 1971, that was a very unusual concept, and, as I said earlier, we needed the confidence of record companies in order to survive.

What Tom saw for The Manor was extraordinary in its time, and perhaps only Rockfield, in Wales, had already put the concept into practice. It all seems strange now, but I recently saw a copy of *Sounds* magazine from 1972 which was making a point of the fact that Morgan Studios, in north London, was then still managing to survive, despite being *8 kilometres* from the centre of London. Recording, in those days, was a very urban industry. At that time, the concept of a studio in the country was not considered to be a serious proposition by the recording establishment. When I announced to my colleagues at Pye that I was moving to a studio in Oxfordshire, 100 km from London, they all thought that I was crazy.

The whole concept of The Manor, a residential studio in the countryside, was, in itself, a very unconventional approach to recording, but when Tom decided to build an 'echo plate' (as the studio could not afford to buy an EMT 140 – the industry standard) … well, my hair loss can be traced back to those days. In fact, his reverberation plate worked tolerably well, despite its somewhat idiosynchratic sound, but when a record company called and asked if we had an 'echo plate', I felt awful saying 'Yes', even though they were not specific about the make. I felt that the make was understood to be EMT, even if unsaid, and that I was deceiving them by not declaring its non-standard nature. It was a very uncomfortable position in which I found myself, because the recording industry was very conservative in those days. Nevertheless, the 'home-made' plate kept us going for about a year, until we could afford 'the real thing'.

Tom Newman was not only a musician, but was also quite a gifted mechanical engineer. He did things to 'improve' the old machines which were unorthodox, to say the least, but my reservations about what he was often doing with them were often based more on my orthodox training rather than the results. I had never worked in such unconventional ways before. By January 1973, I had begun work on The Manor Mobile, as a separate company, with Tom's involvement specifically precluded so that I could get on with doing things 'properly' again, and create a *really* professional reputation for the recording division of Virgin Records, which was a record retailing organisation in those days, and not a record label. During the same period of time, though, Tom was embarking upon a serious re-recording of Mike Oldfield's demo recordings. The working title was 'Opus 1', but it would go on to become the legendary 'Tubular Bells'.

In November 1971, I had begun teaching Tom the basics of recording in a professional manner, yet by mid 1972 he was recording Cat Stevens and Fairport Convention, although still using a good proportion of his 'home-brewed' equipment and techniques. By late 1972, the 'Tubular Bells' recordings were under way, much of which were in a manner that I could not bear to witness. Anyhow, I did what I could to help them over many of their problems, despite, at the time, never believing that the recordings were designed for anything other than a minor, cult niche. In fact, I thought that they would be lucky to sell 5000 copies. So far, it has sold about 25 million.

1.2 Artistic needs

By the end of 1973, The Manor Mobile had established itself as Europe's pre-eminent mobile recording facility, having made, during August, what was believed to have been the world's first live recording by a dedicated, 24-track mobile recording vehicle. Simultaneously Tom Newman had sent 'Tubular Bells' to a virtually resident position at the top of the UK album charts. I had managed to make *my* statement of technical excellence with The Manor Mobile, yet equally, or perhaps more importantly, Tom had made *his* statement on behalf of musicians. Against the odds, he had proved his point, and my respect for him grew enormously. Indeed, there *was* an alternative way of going about the process of recordings. That, I could no longer deny.

At that time, Mike Oldfield was a very insecure character who, to me at least, was impossible to deal with. Tom, on the other hand, had a sort of childish rapport with him that could coax from him some excellent performances which more usual recording circumstances would have stifled. The production credits on the album go to Mike, Tom, and Simon Heyworth, whose combined production credits up to that time totalled to zero. 'Tubular Bells' was perhaps the first really big selling album that was recorded by 'musician power' alone. True, they had the 'workably professional' Manor studio, and adequate technical support from myself, but basically, these guys did it alone. 'Tubular Bells' was not technically all that it could have been, and Mike, himself, was never satisfied with his performances, but none of that seemed to count for much to the record buying public. In 1976, I re-mixed a quadrophonic version for a compilation set, 'Boxed', but plans to replace the original stereo version were eventually scrapped, as it had, by then, achieved 'classic' status. What is more, I never had the original bells for the re-mix. The bells themselves had been damaged by Mike hitting them with coal hammers for some experiment or other. The original recording was no longer on the multitrack tape, and the Belcamman valve compressors, used to get the incredible sustain, had been lost. In fact, it is possible that the tubular bells for the climax of Side One were never on the multitrack, but that they had been 'bounced' directly onto the stereo master. Nobody seems to remember, now, as there was too much Guinness flowing at the time. (Guinness, for those who are not familiar with it, is a famous Irish black beer, and coincidentally, Mike, Tom and myself are all part English and part Irish, though Tom is 25 per cent Russian, as well!)

Mike always dreamed of a 'Tubular Bells' with higher technical standards of both the recording quality and the performance. Given the delicate circumstances of its production, Tom always doubted whether in a situation which could have produced higher technical standards Mike would ever have been out of his psychological shell for long enough to record anything worthwhile. This bugged Mike for years, and was only exorcised by the recording of 'Tubular Bells II', 20 years later, this time with Mike, Tom, and Trevor Horn co-producing. To me, that is a really superb piece of playing and recording. It is virtually flawless, yet for whatever reasons, it will probably never reach the same status of acceptance as the original.

This rather lengthy anecdote is a useful demonstration of how, even at the highest level, conflicts can exist between the requirements for technical accuracy and the needs of the musicians in terms of their performances. Mike

Oldfield now works mainly at home, but his home studio contains the absolute finest equipment. Figures 1.1(a) and (b) show his home set-up in 1975, with equipment that could make mouths water even 25 years later. On the other hand, there are other established artistes who are very comfortable in professional studios, and who keep their homes for their family life and relaxing. For established artistes, with experience and resources behind them, they can choose to record in the circumstances which best suit them. Furthermore, they can afford whatever extra help they may need. The real problem is achieving the first success, and that is where the project studio has been so influential on the recordings of the past 20 years, or so.

1.3 Portable equipment

During most of the 1970s, Tom Newman was pestering me to build him a 16-track tape recorder which would fit into a suitcase. I insisted that the technology was not available, and he insisted that it was, if only I would relax my performance standards. 'I don't need your level of quality', he would say, 'I need to capture ideas, in the right places and at the right times. If the demands of quality prevent such equipment from existing, then an enormous amount of music will never be made.' I am afraid that all of this fell on deaf ears. Lowering standards was an alien concept to me, even if it *was* a means to a greater end. Between 1975 and 1978 I re-built The Manor, and built the Townhouse Studios, in London, both of which were done in conjunction with Tom Hidley, the Californian studio designer. We were pushing the sharp edge of the industry at that time, or so *we* thought. Tom Newman, on the other hand, saw the musical creativity as being the sharp edge of the industry.

Tom Newman's view that the music was the overriding priority is now hard to argue against, but with The Townhouse, we were on the crest of a wave. For me, it was so interesting. It was Europe's first studio with a 32-track analogue tape recorder (Telefunken), and had Britain's first large Solid State Logic (SSL) console, the 4000B. It was one of the first studios to master on ½ inch analogue tape, at 30 ips (inches per second) as its standard format, and the disc cutting rooms were the first to use a tape replay system without an advanced head to control the groove spacing. The direct feed from the normal head fed the information to the groove pitch control, whilst the master audio feed was held in an Ampex ADD1 digital delay. For about a year, the staff were discouraged from speaking to clients about this, as the reputation of digital delays was none too good in those days, but the ADD1 was exceptional. Pushing back these boundaries of technology was exciting for me, and because we had the clientele to match the technology, such as Queen, Page and Plant, Black Sabbath, Frank Zappa, Phil Collins, and many more, I had little time for Tom Newman's persistent requests for his 'toy' multitrack.

For many years, he had 'played' with Teac 4-track ¼ inch machines, but when the Fostex A8, ¼ inch 8-track machine came out, he was one of the first to buy one. Likewise, he was one of the first European purchasers of the later B16, the 16-track, ½ inch, suitcase-sized tape recorder. Finally, Tom had got what he wanted, and he could not have been more pleased. The hours that he worked with that machine, and later with the E16, are beyond calculation. Much of the preparation work for 'Tubular Bells II' was done on these

Figure 1.1 (a) and (b) A home studio (though very elaborate for its time) in the house of Mike Oldfield, on Hergest Ridge, bordering England and Wales. In 1975, this set-up was used to record and mix his 'Ommadawn' album. The equipment includes a Neve mixing console, an Ampex MM1100 24-track tape recorder, DBX type I noise reduction, and a Westlake Audio monitoring system.

machines, with various small mixing consoles, and although, in my opinion, the sound quality lacked the sort of naturalness and transparency that I associate with professional systems, Tom recorded several albums on these set-ups that were subsequently released on CD; not as re-issues from vinyl originals, but as original CD releases. Such a set-up is shown in Figure 1.2.

Remember, though, that Mike Oldfield had never been satisfied with the recorded quality of 'Tubular Bells', so I still felt justified in trying to maintain professional recording standards, even though by then I fully accepted the viability of their alternative approach to recording. If the 16-track in a suitcase was not going to produce the quality that I was brought up to expect, then I still resisted having standards pulled backwards, because I saw it as the thin end of a wedge which could ultimately de-value all the recording standards. The then typical home recording set-up was still a long way from producing state-of-the-art results.

I remember the 'buzz' around the industry when it was revealed that Eurythmics had recorded their big selling 'Sweet Dreams' album on a Tascam 8-track, or something similar. It is true that, by the early 1980s, Japanese miniaturisation of recording technology, together with some significant advances in the formulation of recording tapes, such as Ampex 'Grand Master' 456, had made possible recording machines which only 5 years earlier would have been merely the stuff of dreams, but, nevertheless, top-of-the-line equipment had also advanced. Furthermore, professional digital multitrack machines had arrived. What was possible in the finest studios was still well ahead of what could be achieved by the 'semi-professional' systems.

However, questions were being asked such as 'If a well-received album

Figure 1.2 The writing/pre-production studio of Tom Newman, in 1991, where much preparation work was done in advance of the recording of Mike Oldfield's 'Tubular Bells II'

Figure 1.3 Sun Zoom Spark (Sunsonic). This studio was built in Shoreditch, London, UK, as a very temporary set-up in 1989. It was built in an old warehouse for a group who had signed a deal for recording one album, with the record company having an option on two more. Despite being built in a rather rudimentary manner, and very quickly, it still exists, re-equipped and under new ownership, 10 years later

such as 'Sweet Dreams' could be made on semi-professional equipment, then did it really require the finest professional equipment to record the Sex Pistols rendition of 'Pretty Vacant'?' Personally, I would say that all recordings should be made on the best equipment available to them, but the Tom Newmans of this world would also say that no piece of music should ever be prevented from existing, solely because the best professional equipment was not available for its recording. The music comes first, that I accept, but I still believe that one should record it with the best facilities available for the recording. Sound quality, alone, should never be the absolute determinant, but neither should it be subjugated excessively to convenience. If one is dealing with a new artiste, with no record company support, then one does what one can. On the other hand, we have an elite side of the industry, which is only interested in working to the highest achievable standards. In truth, both are necessary, because both lead to advances in their own areas. When they can be brought together, they lead to a general advance for the whole artform/industry combination. However, even in the late 1980s, the technology gap still entered the equation all too frequently.

1.4 The project studios

By 1990, as was the case with many, if not most, other studio designers, I found myself in a situation where an increasing proportion of the studios which I was being asked to build were for private ownership and use, with some also being for a limited amount of commercial use. Although the quality gap still existed between the semi-professional and top-of-the-line equipment, it was often deemed that the semi-pro equipment was not 'good enough' for many purposes. I found myself building studios, some perhaps for only a three-album project, and with only a limited intended life. These really *were* project studios, built for one project, or one set of projects, only. One such studio is shown in Figure 1.3, though it way outlived its intended life-span. Some clients wanted a more long-term approach, and opted for excellent rooms which would initially house 'budget' equipment, but which would be worthy of better equipment when it could be afforded. Others wanted to invest mainly in equipment, with rooms that could be up-graded when it could be afforded. On balance, the first approach seemed to work the best, because good rooms meant good sounds entering the microphones, and good monitoring conditions allowed the best compromise decision to be made when assessing the sounds from the budget equipment. Good equipment, on the other hand, can rarely achieve the best of its capabilities with strange sounds entering the microphones, and with monitoring conditions which left the engineers 'driving through fog'. However, on different types of music, and with talented people, some good results emerged from both approaches.

The upsurge in the building of these types of studios was driven by two developments. Firstly, a huge proportion of the then current popular music was being based around the recording of electronic, keyboard controlled instruments. These were largely directly injected into the mixing consoles, so the need for many expensive microphones and good recording rooms was reduced. Secondly, realising that this new approach to recording was largely electronically based, the record companies began advancing money to the artistes specifically for the purchase of the necessary equipment to make such recordings. As equipment prices were seemingly ever-reducing, in real terms, this option became financially attractive. The final cost of the recording was known at the outset, as there was no problem with extensions of the expensive studio bookings due to schedules falling behind. Unfortunately, all too often, neither the record company accountants nor the artistes really knew many of the subtleties involved, and equipment vendors had a field-day selling equipment, much of which was really not as good as the salespeople were claiming. Poor monitoring in acoustically untreated rooms also did nothing to help the assessment of what the audio quality was really like. The pro/semi-pro quality gap still remained, but many recordings sold well despite their various short-comings. The trend for this type of deal soon became established.

Figures 1.4 and 1.5 show yet another approach, where the owners opted for good basic acoustics and monitoring, though in some rather unusual buildings (unused parts of other premises which they owned), in combination with old, secondhand, but highly professional equipment. There is much to be said in favour of this option, though 'easy terms' payments may not be available. Sound quality is usually beyond reproach, interfacing can be done with full, professional simplicity, and wide operational tolerances can be very forgiving of inexperienced operators.

Figure 1.4 (a, b) A private studio in St Albans, UK. (a) The space created by the installation of a mezzanine floor in an engineering (metal spinning) factory. The high-density foam isolation blocks can be seen under the first layer of the sandwich-type floor construction

Figure 1.4 *cont.*(b) The finished control room, and the glass doors leading to the studio. The mixing console is a re-furbished Cadac, from 1972, bought for only £2000 sterling

Figure 1.5 (a, b, c) The Gazebo (Silvermere Sound), Ascot, UK. (a) A summer-house in the garden of the home of a record producer

1.5 The digital age

Suddenly, around 1994, a revolution took place. The arrival of some inexpensive condenser microphones from eastern European countries and the former Soviet Union, together with mixing consoles such as the Mackies, and the 8-track modular digital recording systems, notably as the ADATs and DA88s, opened up a whole new level of achievable quality for the amount of money involved. If special facilities are ignored, then the difference in the basic recording quality achievable from systems made from the above components, when compared to that when using Neumann microphones, Neve mixing consoles, and Sony 3324S tape recorders, is not very much. Given good monitoring conditions during recording, when due attention can be given to the sonic subtleties of the process, the two systems can yield recordings which can be hard to tell apart. Even many experts would find difficulty in deciding which one was recorded on which system.

The greatest quality gap which currently exists between the project studios and the main line studios is now largely in the domain of a knowledge gap, between the acquired skills and experiences of the true professionals, and the operators of many of the budget studios. In fact, it is not only the operating knowledge which is important, but also the interfacing knowledge, because much equipment marketed towards project studios is very lacking in standardisation of interconnections, as we shall see in later chapters.

The project studio business has now become a huge international market for equipment suppliers, and market forces tend to be quite ruthless. Things are marketed as being easy to install and easy to use, but, in reality, this is

Figure 1.5 *cont.* (b) The inside of the summer-house after conversion into a control room. The main monitors are early Questeds, and the mixing console is a Trident, Series 70

Figure 1.5 *cont.* (c) A side view. The chimney has been pierced by a window, giving a view into the wooden extension that can be seen to the left of the chimney in (a). This was transformed into a small vocal room. (The wooden extension, that is, not the chimney.)

often not the case in many typical circumstances of use. Blind faith in the advertisements and magazine reviews can lead to some very disappointing results, and due to the number of cut-backs to keep the prices competitive, the interface of much 'semi-pro' or 'promestic' equipment can be much more problematical than for their professional equivalents.

These days, things are often marketed as 'black boxes', with simple inputs and outputs, yet, in many cases, a knowledge of the principles of operation is still necessary if pitfalls are to be avoided and the best results of their use are to be achieved. There is no longer any need to accept the sort of performance limitations which were inherent in the semi-professional equipment of 10 or 15 years ago. What many project studios now have, in terms of equipment, is capable of very fine quality recording. It now seems to me to be an absolute waste when so many limitations in the end results stem from ignorance, and ignorance only. I have personally made recordings, the quality of which have been superb, on some of the current 'project' type equipment, so I know from first hand experience what quality of recording it is capable of achieving.

In the days when Tom Newman was asking me for a limited performance 16-track machine in a suitcase, a quality limitation for the audio was deemed by him to be an acceptable compromise if it led to the advent of good music, which would otherwise never get the chance to manifest itself. From my own point of view, though, I could not get very enthusiastic about down-grading my standards, but there again, I was working in a different part of the industry, mainly working with already established artistes. The point to be made here, though, is that such significant performance level differences no longer exist.

Listening to CD releases of classic albums from 25 years ago, such as 'Tubular Bells', or the Pink Floyd's 'Dark Side of the Moon' , on a good hi-fi system, will reveal many recording defects which, by today's standards, would be generally considered to be unacceptable. Nevertheless, what all of these classic recordings have in common is that the people involved did their best, using equipment which was optimally interfaced and well maintained. The question as to whether they were good enough for release did not arise, because they could do no better. With current promestic equipment, the performance possibilities are really quite incredible, yet on my travels, I see so many recordings being made under circumstances where their true potential is by no means being fully realised. In many cases, I see so many of the most basic principles of the recording process being violated, usually in the misguided belief that a chain of 'black boxes' will behave as the sum of its individual parts, even if they do not interface too easily. Nowadays, it all too often seems to be the case that how far to bother with the niceties of what *can* be achieved is more a question of what other people in similar situations are achieving, as opposed to what one *should* be striving for. This is, perhaps, somewhat more a question of marketing policy than an engineering one, but just how far should we go in attempting to get the best out of a project studio which may only be producing low cost products? Well, I believe that it is a travesty when *any* recording fails to realise its full potential, and it is especially so when the limiting factors are down to a lack of education. In fairness, though, it is not all that easy for many users of project studios to find the references that they need, and manufacturers are often disgracefully lax, or even downright deceitful, in their marketing approaches, so it is little wonder that many project studios operate in a mist of confusion. So, what the following chapters will seek to do is to throw some light on the commonly encountered problems, and, where necessary, bring the spotlight on the manufacturers' practices which regularly let down their customers. First, however, we should take a careful look at some of the situations which exist, in order to get a general feel for the reasons why some of the more specific, detailed discussions will be necessary.

Chapter 2

The equipment and the compromises

A substantial proportion of this chapter will be rather philosophical, but without an adequate understanding of many of the realities of the current recording world, the relevance of what is discussed in some of the later chapters may be missed. The previous chapter looked at some examples of how we have come to arrive at our present state of affairs, and this chapter will discuss something of the circumstances in which we now find ourselves. Project studios have come into existence as a result of market forces, and those forces are, to a great extent, still their masters. These forces do not always have high fidelity at their roots, though, so it is better to appreciate what drives them if the best results are to be extracted from where they take us. By now, the equipment available to the project studio users exhibits an enormous range of quality, which stretches from the fully professional down to the most basic of 'Porta-Studio' type devices. There are also many basic studios, these days, which possess the odd item(s) of unquestionably professional hardware, whilst, conversely, some items of very arbitrary quality have found their way into professional use, though mainly for special effects. This again highlights how the pro/domestic (promestic) divide has closed.

2.1 DATs, MiniDiscs and NS10s

The DAT (Digital Audio Tape) and MiniDisc formats were both intended for the domestic market, and Sony fought very hard, in the early days, to emphasise the fact that they were not conceived as professional formats. Nonetheless, the major studios began to use DAT machines in great quantities, and much work was put in by the various manufacturers to make the system more robust. Personally, I still keep back-up tapes of anything which I have on DAT, as the sudden appearance of 'glitches' is all too frequent, and I would never feel secure if I had any 'irreplaceable' recording solely entrusted to one DAT cassette. As a domestic format, it never really gained much acceptance, but it has now become a well established standard format in professional use despite its fragility.

The story of MiniDisc is largely the same. Here, robustness is not the limiting factor, but data compression. A device which has so far failed to be widely accepted by the general public has found a definite pair of markets in the professional world. Many classical recordists use the MiniDisc as a second-line recording system, which they can use to make experimental edits. By this means, they can quickly check whether the edits will work on the master tapes, or whether a further recording is needed. The editing functions of MiniDisc are excellent in that they are both quick and simple to achieve. The efficiency of their on-board editing capability has also led to their widespread use by the broadcasting industry. For such purposes as news-gathering, the data compression is not a limiting factor, as the quality is still excellent by comparison to many other news-gathering recording systems, not to mention many of the subsequent transmission/reception systems.

The Yamaha NS10 loudspeaker is yet another example of a piece of equipment, purpose-designed for the domestic market, which was widely adopted by the professionals. Again, as a domestic product, it failed to live up to its expected sales potential. The early NS10s were prone to regular failure in professional use, and were deemed by many to be too bright sounding; hence the sheet(s) of toilet paper which were regularly seen covering the tweeters. Whether one sheet, or two, and of which brand, was the subject of endless debate. In order to correct these problems, and in recognition of their professional usage, Yamaha brought out the more powerful NS10M, which again, despite some initial comments about its sonic dissimilarity to the original, was soon widely accepted in professional circles.

That these things were never launched as professional products is due to the fact that the manufacturers recognised that they all had shortcomings that professional equipment should not have. Robustness in daily use, and a reliability that will continue for year after year, are usual prerequisites for fully professional audio equipment. When DAT was launched, its longevity was unknown. Huge amounts of money are involved in the research and development of such equipment, and commercial realities dictate that these things must be launched into the marketplace before they are fully proven. If a competitor steals a market lead, due to delays, it is quite possible that the R&D costs may not be recovered, which is bad business. Sony realised that if this medium had been launched as a professional product, they could have been liable for huge financial damage claims if the things failed to perform either in a professional manner, *or*, for an amount of time which could not be predicted. Nevertheless, the DAT fell into an awaiting hole in the list of equipment needs, and the rapidity of its general acceptance has been phenomenal, even if there are a few die-hards who, perhaps quite rightly, remain cautious about DAT.

The MiniDisc was never intended for professional use because of the data compression systems, which had to be used in order to get so much information on such a small disc. Truly enormous amounts of money continue to be spent on research into the audible limits of data compression, but, as yet, no absolute conclusions have been reached in terms of how much compression can be used before a system loses its audiophile status over a wide range of musical programme material. The 16-bit linear, 44.1 kHz sampling rate of the compact disc (CD) format is already widely recognised as *not* being adequate for the highest end of hi-fi. When CD was launched, in the early 1980s, it was close to the limit of the available technology, but advances have now yielded 24-bit linear, 96 kHz systems, and even 192 kHz systems are now appearing. Twenty-one-bit, linear, 96 kHz would seem to be a good compromise between extreme fidelity and commercial realism.

However, it should be realised that if the current CD format is no longer considered to be sufficient, then any system which uses data compression with performance limits below that of CD *cannot*, no matter how clever the psycho-acoustic masking algorithms might be, better a 16-bit linear, 44.1 kHz CD. Therefore, anything which is mastered on to a 'less than CD' format can never be upgraded in the future. Also, it seems that there will always be some types of music signals which will render any form of lossy data compression to be audible. Whether the audible artefacts actually render the music less pleasing to listen to is a moot point, but the general professional attitude is that high quality music recording should be mastered on the available medium with the fewest sonic limitations.

The NS10 is a freak. As a hi-fi loudspeaker, it is not very good at all. It has a coloured response, a harsh mid range, a peculiar off-axis response, poor resolution of fine detail, an inadequate overall frequency response, ... the list could continue. I doubt that anybody would have dared to launch such a product specifically as a professional monitor. Despite all this, if I am asked to check out a multi-way recording studio monitor system, and have no access to any measuring equipment, then I can set the systems to within a very small deviation from optimum if I have a pair of NS10s as a reference. I do not mean that I set up the large monitors to sound like big NS10s, but rather that from the NS10s, I can judge, quite accurately, what the other loudspeakers *should* be sounding like. The NS10 is a very useful industry reference. Almost every professional knows what it sounds like, and it does possess a certain something, which can be very hard to define, but which seems to enable relative balances of instruments to be judged with ease. I find them to be good loudspeakers on which to get a mix going, but I would never work with them alone, as they leave too much unheard in terms of noises, phase problems, and things at the extremes of the frequency range. These points will be discussed further in Chapter 5.

The above three products all found their way into professional use, despite their shortcomings, as they each fulfilled a purpose in areas that were deemed to be wanting. When they are used professionally, however, they are used with a full knowledge of their limitations (well, *usually*, at least) and in an appropriate place within a professional set-up. An error sometimes made by the owners and operators of project studios is to see these things as being fully professional because of their widespread use within professional operations. To record on to MiniDisc, edit within the machine, and then to transfer the recording to DAT, for archiving, all whilst monitoring via NS10s, is *not* a professional approach to high fidelity music recording. The sonic quality of the MiniDisc would limit the achievable performance from the DAT copy, the DAT would not provide the robustness and reliability for guaranteed, later retrieval, and the NS10s would be unlikely to show up many of the potential editing artefacts, or low level distortions or coloration in the sound.

It would be an easy system to use, however, and the results may be adequate for the music being recorded, but if such results are deemed to be acceptable to you, then you need read no further, as the thrust of this book is to get the maximum performance out of the equipment which is now readily available, rather than accepting 'good enough'. This does raise some interesting philosophical points, though, which may never have any absolute conclusions. The pursuit of 'quality at any cost' is not realistically viable, especially in the context of this book, because we are discussing project studios. Nevertheless, the professional attitude is surely to do the best that one can under the circumstances. Understandably though, the commercial attitude may strike a lower quality point of compromise. The following section is the text of an article written for the March 1997 *Studio Sound* magazine[1], which may help to highlight my own views on this subject.

2.2 Good enough?

In the 'Open Mic' column of the November 1996 *Studio Sound*, Martin Polon spoke out strongly about the way in which the music business has largely settled for eroded standards in the race for profit. 'It is time to stop this, and make audio quality our byword,' he stated. I couldn't agree more, but it will be an awfully uphill struggle to do so. Not only have economic recessions forced cut-backs, but also, falling equipment prices, in terms of value for money, have made it possible to provide recording facilities in a garage, or a garden shed, which incorporate things which even 15 years ago would never have been dreamt of. On the other hand, they lack a lot of facilities which were once considered to be essential. In reality, most of these things probably still *are* essential if top quality recordings are being sought, but, once standards are relaxed, it can be very difficult indeed to re-establish them. What is more, there are whole industries which are now geared up to maintaining the status quo, and a sort of Mafia of Mediocrity is taking hold.

The power of manufacturers is enormous, and it grows from day to day. The bulk of recording equipment is now made by multi-national conglomerates, and like most large companies, they exist for one reason – profit! When companies become listed on stock exchanges, their shares are purchased by investors, most of whom have little or no knowledge of what goes on inside the industries in which they are investing. They invest to make money, and companies are there primarily to serve their shareholders, *not* their customers. If writing the word 'professional' on equipment makes it sell more units, then manufacturers will do so, whether the equipment *is* professional or not. I know of numerous companies who have ostensibly sold 'professional' equipment, without providing anything like the customer support that professional users would require. If I have a problem and I call a truly professional company, they are 100 per cent on the case, trying to help, yet I have been involved in hugely expensive live recording situations, using notionally 'professional' equipment, where calls to manufacturers have resulted in 'You must deal with your local agent. It is not our company policy to discuss these things directly.' This could be on a Friday night, in a far-flung location, where the dealer would not be open until the Monday morning, and even then, would probably have to refer back to the manufacturer, perhaps with difficulty in a foreign language. Too little, too late; this is not professional. [Chapter 6 will relate a true horror story to which the previous few sentences refer.] Only recently, when I needed back-up information from a 'legendary' company, who had not too long ago been taken over by one of the huge conglomerates, I kept getting routed to a service department answering machine, and nobody called me back after six attempts over two days. Ten years ago, this same company would have moved Heaven and Earth to help me out.

Record companies also want profits, and most of them *will*, make no mistake about it, work to the minimum standard (cost) of recording that will yield the best profit. If something is too badly produced, it won't sell, but if it is too expensively produced, then doubling the recording cost may only add 1 per cent to the sales. This is seen as bad business, and the shareholders will not like it. Of course, I'm not talking about the Telarcs or Deccas of this world, who sell quality, or the Peter Gabriels and Pink Floyds, who work to their artistic limits, but they are only the smallest tip of a huge international industry. In *many* countries, the fast cars and large houses of the company executives come before the recorded quality of the music. It is business, pure and simple.

Along with the rise of the 'project' studios has come a huge increase in the power of advertising. I remember discussing this subject with the chairman of a well-known company making quite expensive mixing consoles. He told me that he did not remember a single sale coming directly from an advertisement. 'I advertise on behalf of my customers' he said. 'However, if we don't advertise, hardly anybody will buy our consoles, because *their* clients will not be likely to book a studio with a mixing desk that they have not seen in the magazines.' Much of this is driven by an industry that is expanding faster than it can properly train staff, and so much happens as the result of hearsay, rather than experience. Small manufacturers of audio products have a battle on their hands, trying to survive in a publicity dominated world. The cost of maintaining an international advertising campaign is enormous, and can only be supported by 'mass production' manufacturers. It cannot possibly be paid for out of low quantity sales. The sad thing is that the small, specialist manufacturers were at the heart of the quality development. I was sad to hear of the demise of UREI, its sales being insignificantly small in the eyes of its enormous new owners, to whom producing only a thousand units a year of a particular product was of no significant interest.

Another problem is the music/recording press. Magazines like *Studio Sound* are now all too rare, where open, uncensored, experienced debate can be undertaken. Many magazines now aim at the mass market sales, never publishing a derogatory equipment review, and only publishing articles of lengths suitable for people with only minimum attention spans. Furthermore, in some countries not in the 'first division' of the recording world, I know of magazines where they do not pay for articles, but will only publish an article if the writer, or writer's company, buys advertising space in the same, or a subsequent, issue. Dealers thus finance articles about the products which they sell. (Incidentally, I am not referring to small, Third World countries here, but large, European countries, including ones inside the EU.)

In a world of blind people, a one-eyed person will be the ruler. Many dealers are seen, by their association with the big names which they sell, to be 'knowledgeable', and frequently are very defensive of their positions. A two-eyed person, perhaps seeking to publicise some uncomfortable realities, may be prevented from writing in the magazines by threats of withdrawn advertising if they try to give new eyes to the blind. There are dealer 'Mafias' in existence which hold great sway over the press, and hold back the development of whole national recording industries. Just imagine that you cannot read the international, English language Press, because you speak only Spanish, or perhaps even worse, Greek, which is in less international use.

Where do you get your latest information from? There is little enough Spanish language recording press, and some international magazines which have Spanish versions are not translations of the original language issues. The content can be very inferior.

All round the world, there are dedicated and experienced people, such as Martin Polon, who care about what is happening. There also exists an elite, international recording industry, which is perhaps the 'university' of recording. As Dan Daley was saying, though, in his October 1996 column of *Studio Sound*, 'There is little growth at the top, and the business is where the growth is.' Unfortunately, however, this expanding section is the one most vulnerable to the power of advertising, rather than reason, and in this world, which sees the 'super-studios' as being in another dimension, I fear that the worst of Martin Polon's worries have already become a reality ... 'Good enough *is* OK.' It is peer standards, not absolute standards, which determine what is good enough for the bulk of the recording industry, and there are whole marketing industries trying to keep it that way.

2.3 Commercial or professional?

Well, notwithstanding the contents of the previous section, we will continue looking at how to get the best results that we can, because most, in fact virtually *all*, of the professionals who I know are always seeking to improve their standards. This is one of the fundamental differences between a professional and a business person, the professionals always seek to improve their standards, and they gain satisfaction from the results of their skills. The business people tend to count only the profits. Now, we must be realistic, and often the business people make it possible for the professionals to survive, so *their* needs must also be respected, but for the purposes of this book, we will concentrate on the improvement of standards. The compromises to commerce we will have to leave to the individual readers.

Before we leave the subject of commercialisation, though, it may be worth considering another observation of commercial trends, because unless we can get this subject fully understood, it may be difficult for some readers to establish the free-thinking which is necessary to be able to carefully balance their requirements and be able to achieve the best results for their own purposes. The typical equipment package from a dealer may not be the best way to go, but the pressures and temptation to accept such a package may sometimes be difficult to resist.

2.4 A question of balance

When clients commission me to design a studio, then they engage my services as an adviser on how to achieve a complete recording system. Dealers, on the other hand, are often desperately trying to influence them in other directions. The majority, it would seem, do not care one jot about *my* problems of building the integrated system which I believe to be in the best interests of my clients. Head-on conflicts between designers and dealers seem to be on the increase. Beware of the dealer who six months ago told you that the

Figure 2.1 (a, b, c) Project studio for a small record company (Revolver Records, Wolverhampton, UK). (a) The control room, built in what was the front lounge of a house. The glass doors lead to an air-conditioned machine room, converted from the front window alcove. The console is an Amek Mozart, and the monitor loudspeakers are Tannoy; plus, of course, the Yamaha NS10s

products of 'manufacturer A', which they then represented, were the best, but now that they have either lost the dealership or changed their allegiance to the more profitable range of 'manufacturer B', say that manufacturer A's products were really not very good. Beware also the dealers who always want to sell the next most expensive unit to the one which the designer has specified. The designer will most likely be thinking about the best overall end result, the dealers will more likely be thinking only of the profits.

It is a tough world out there. Manufacturers put pressure on dealers to sell a minimum amount of product each year, or risk losing their dealerships. Remember that fact if you find yourself confronted by dealers who try to sell you not what you were asking for, but what *they* want to sell you. This is especially true if they seem to be treating your initial requests with disinterest, or trying to make you feel ignorant. *Rarely* take the advice of dealers over that of designers. Designers are probably considering a balanced system. It seems to be intuitively obvious to Ferrari owners that if a Porsche agent tries to sell them a new-style Porsche gearbox, it is unlikely that it will fit comfortably into a Ferrari, for which it was not designed. Unfortunately, however, it does not seem so obvious to many studio owners that much equipment may not be suitable for an *environment* for which it was not designed. There is a question of balance which extends to each differently designed recording system. Balance, in the recording industry, is not exclusively an issue of the relative levels of musical instruments.

Figure 2.1 *cont.* (b) and (c) The studio, built in what was formerly the rear lounge. The photographs were taken from the live end of the room, looking towards the variable end, where hinged hard/soft panels can be moved to expose hard and soft surfaces as required. The triangular wall protrusions, at the junction of the two halves, house curtains that can be drawn across to divide the room further

Many absurdities happen due to the insecurity of many studio owners and operators about their beliefs in their own ears and musical judgements, and it is on these insecurities that many parasitic operations thrive. The only real solution to the insecurity problem is experience and education, and hopefully, via the pages of a responsible press, more of this can be shared out, which will hopefully lead to a healthier recording industry; and that is something which I care about greatly.

2.5 Domestic, semi-professional or professional?

Domestic equipment is primarily designed for music *re*production, although there *is* a limited amount of dedicated 'home recording' equipment. The only purely domestic pieces of equipment which are likely to find their way into project studios are amplifiers and loudspeakers, though cassette machines do find use for domestic reference, from time to time. Domestic equipment tends to be of a light duty nature, and is normally designed to handle complete programme material. The presumptions are also usually made that low level interfacing will be via short cables, and that the system complexity will not be very great. Signal levels between different pieces of equipment tend to run about the 300 mV level, or, in other annotations, approximately –10 dBV, or –10 dBu.

Figure 2.2 September Sound, London, UK. The studio in which, for many years, the Cocteau Twins recorded many of their albums. Although not an ideal acoustic situation, they considered the view across the River Thames to be of prime importance. The studio was originally built on the top floor of Eel Pie Studios, owned by Peter Townsend of The Who, but September Sound now operate the whole facility. This photograph was taken in 1989, and the operation has expanded considerably since then.

Figure 2.3 A tiny control room with excellent acoustics built in an attic with less than 14 m^2 of total floor area. 'Noites Longas', near Seixal, Portugal, was built as a private studio, but was so successful that it now operates as a full-time commercial venture. The mixing console is by Mackie, and the monitors are Reflexion Arts, plus, once again, NSIOs.

Professional equipment, despite often having very similar sonic performance to the better domestic equipment, has a very different set of design criteria. Professional equipment needs to run, reliably, for long periods of time, with its daily usage being at a very high level. Professional systems can be complex in their levels of interconnections, with many parallel and serial connections existing at any given time. To make such interfacing easier, signals are run at much higher average levels than those of domestic equipment, with typical levels being 1.23 V, or +4 dBV. Incidentally, dBu is the voltage level relative to the voltage that would have been necessary to drive the old standard of 1 milliwatt into 600 ohms, or 0.775 volts. dBV, on the other hand, is the voltage level relative to 1 volt, but these matters will be discussed further in Chapter 3. Suffice it to say here that they all have good reasons for existing. Large recording sessions can be very expensive affairs, so reliability and relative immunity to interference from external sources are subjects of great importance when it comes to professional equipment design. The balancing of low level signals, such as from microphones, *and* line level signals becomes very important in professional operations.

Unfortunately, these professional requirements tend to be expensive if they are not only to do their jobs, but are also to perform to the highest sonic standards. One of the main restrictions on the performance of semi-professional equipment is cost, as such equipment needs to provide the sonic performance of the better domestic equipment with the interfacing needs of the professional equipment. In short, for the prices that such equipment can support, it cannot usually be optimally achieved, so compromises must be made. These usually take the form of less sturdy construction, less flexibility in use, and smaller windows of optimum performance. Serviceability may also be reduced, such that routine servicing is less easily accomplished and may take more time to complete. In many circumstances, excellent results *can* be achieved when recording on semi-professional equipment, but more care may need to be taken. The sonic performance of such equipment in skilled hands may deliver results which, on many types of music, may be virtually indistinguishable from those recorded on fully professional equipment. However, due to the tightness of the optimum performance windows, more skill and experience may be needed to achieve the best results from semi-professional equipment than from a fully professional set-up. Perverse, isn't it?

In the case of analogue tape recorders, this limitation can be more critical than with digital recorders, but it also relates to many mixing consoles, not all of which have true sonic purity over their entire, specified dynamic range. When I build large, professional mobile recording vehicles, one of my prime specification criteria has always been adequate headroom throughout the whole system. (Readers unfamiliar with the concept of headroom should refer to the glossary.) Sound checks for live performances, and in studios, too, for that matter, rarely produce the sound pressure levels which the actual performances achieve, so a wide window of acceptable performance from the recording equipment is a must. If the operational window between unacceptable noise and unacceptable distortion is too narrow, then the anticipated level settings may not allow for unexpected peaks to pass cleanly through the system. Setting initial levels too cautiously may lead to recordings having less than optimal noise performance. In a fixed studio, as opposed to the recording of live performances, the wider window is not so critical in terms of catching a one-off performance, but even in the studio, re-takes waste time, and musicians can easily 'go off-the-boil', so even here, the flexibility of operation enjoyed by professional equipment is a great benefit.

However, another limitation of semi-professional equipment, especially of the lower end of the performance range, is the marginality of the sonic performance. In many instances, each component part of a chain can achieve adequate sonic performance in isolation, but the marginality of each piece can render the whole chain to be less than optimal in sonic terms. Some of this can be down to interfacing inconsistencies, but the effect can also be demonstrated by comparisons of different tape recorders. If we were to take two tape recorders of the same brand, one from the top of the professional range, and another from their semi-professional range, we could make some interesting comparisons. When making simultaneous recordings on both machines, each accurately aligned and recording at optimal level, even experts may have difficulty in telling which recording was which in blind tests. On the other hand, if the recordings were bounced from track to track, say four times on each machine, then the quality of the professional machines on the fourth generation would be more likely to be recognisably better. What is more, if the tests were repeated, both at 5 dB above optimal levels and 5 dB below, the fully professional machine would be even more likely to show its pedigree. Track bouncing at these levels would soon show up the deficiencies of the cheaper systems.

Once it is understood that the metering differences can be so very signal dependent, even between typical analogue machines (a subject dealt with at greater length in Chapter 7), the ability to find the precise optimal recording level for any given instrument can be a very difficult task. Again, when rapid working is necessary, a wider window of high quality response is a definite asset. There are many times in top line studios when a piece of semi-pro equipment is auditioned, and people are left wondering exactly why they should pay four or five times more for the professional version, which seems to sound the same. Some of the above discussion will hopefully have helped to explain why, but there is yet another facet of professionalism which is essential for front line use, and that is maintenance back-up.

2.5.1 The professional necessities

When one sees the word 'professional' written on the front of a piece of recording equipment, it immediately leads many professionals to suspect its validity. The vast majority of truly professional equipment is obviously so, and therefore does not require any label. 'Professional', however, is a magic marketing word. It appeared on the front of the first Alesis ADAT machines, but any attempts to get professional back-up from the factory would soon reveal just how amateurish the operation can be. With these machines, somewhat infuriatingly, there were no servicing instructions for the users, no help available from the factory direct to the customers, and in many countries, the importers, who act as service agents, were so inept that they have frequently been known to render machines to all intents and purposes inoperable, by aligning them in a way in which they would no longer play tapes from other machines. More of this later.

Whilst on the subject of alignment, though, it is horrifying to see how many project studios do not even possess any alignment tapes for their analogue tape recorders. Not only do different brands and different types of tape show variations in their required alignment, but even different batches of the same type can show response variations. If machines are not aligned as closely as possible for the tape in use, then the errors will accumulate when copying or track bouncing. In domestic circumstances, perhaps it is wise to leave such things alone, as by far the majority of purchasers will have no knowledge of how or why to make the adjustments, but there comes a point where the further advancement of many project studios becomes restricted by the lack of control over machine alignment parameters. This fact even extends to digital machines.

In the case of fully professional equipment, the manuals are usually very explicit about the alignment procedures to follow, and the necessary controls are usually readily accessible. It is true that there is a potential for mis-alignment in inexperienced hands, but this is something more of an issue for domestic, or consumer equipment. Once a company begins to market something as professional, or even semi-professional, they are beginning to take a rather patronising attitude if they still want to 'idiot-proof' their equipment. One of the very fundamental requirements of machines for professional use is that they should be field serviceable, with fully descriptive servicing manuals and readily accessible adjustments.

Despite its complexity and a price of about $160,000, a Sony 3348 48-track professional digital tape recorder *is* field serviceable. Each one is supplied with a very in-depth maintenance manual, and whilst one would have to be rather stupid to begin poking about inside such a machine without the required knowledge, the means to do so is provided for anybody who *can* understand the manual. The machine does not sport the word 'professional' anywhere on its frame; it merely is. I must say that the reliability of these machines, and their alignment stability, even in arduous conditions of use, is almost beyond my comprehension. The odd thing is that although they may appear to be expensive to buy, over 10 years use, or so, they tend to work out cheaper than the buying, the sending away for repair, and the regular replacing of half a dozen modular 8-track machines, even when adding back in their resale value. More to the point, this does not even begin to take into account the cost of lost time due to alignment and synchronisation problems with the modular machines; many of which do sport the word 'professional', but are patently not.

Semi-professional equipment in limited set-ups can perform very well, but as I have indicated, if a system is built up from as many channels, tracks, effects, processors, and whatever else that would be found in a fully professional studio, then it is doubtful that noise, distortion, high frequency compression, or general sonic neutrality could reasonably be expected to match the performance of a similar chain in a fully pro studio. This fact simply does not seem to have registered with a large number of project studio users. In short the differences between the top and bottom of the range machines are that the top of the range machines have a wider operational window between noise and distortion or compression, and their sonic neutrality is usually also greater. The machines will also last longer in everyday use before their performance degrades to an extent that they become inadequate for their function.

So, there are some unquestionable restrictions of operational flexibility which may be imposed by some lesser specified equipment, but with good monitoring conditions *and* timely reference back to a first generation of recording, any build-up of sonic degradation can be assessed, and steps can be taken to prevent its intrusion into the recorded quality. As in most other walks of life, an awareness of the pitfalls is a great asset to progress. But, take a chain of each level of equipment, and transfer recordings four times through the chains, and any of their discrepancies should begin to become obvious. This is not an unreasonable exercise, as recordings pass through consoles on initial recording and again during mix-down, and perhaps again when tracks are bounced or sub-mixed. On the fourth generation of the process the sound of most non-professional chains will almost certainly be significantly degraded by comparison to the fourth generation from a fully professional chain. This holds true for a great deal of 'domestic'/'semi-professional' equipment in general. The major implication of this fact is not to render many semi-professional recording chains to be inadequate for high quality recordings, but that they should be used sparingly. Do not pass through any unnecessary amount of equipment; and be aware of the degree of any cumulative performance limitations.

2.6 Which way to go?

There are many approaches to the designs of project studios and the choice of equipment. Some possibilities are shown in the photographs distributed through this chapter. Figure 2.1 shows a studio in Wolverhampton, England, where a very respectable range of equipment has been installed in two adequately designed rooms on the ground floor of a semi-detached house, owned by a small record company. Figure 2.2 shows an entirely different approach, where atmosphere was all-important. The inspiration of making music whilst overlooking the River Thames, in London, was considered by its owners, the Cocteau Twins, to be fundamental in their approach to recording. Both of the above studios use fully professional equipment. By contrast, Figure 2.3 shows a studio in the attic of a house near Seixal, in Portugal. In this case, the owners opted for the use of well chosen budget equipment, but installed it in an acoustically controlled room with excellent monitoring, which ensured that when any sonic limitations were encountered, they were immediately recognised, and could be dealt with before they became a problem.

The control rooms shown in Figures 2.1(a) and 2.3 were for use by freelance engineers as well as the studios' own staff. The control room in Figure 2.2, on the other hand, was primarily for the use of its owners, who had time to get used to the idiosyncrasies of a less-controlled room acoustic. All three rooms have produced excellent recording. However, the equipment from Figure 2.3 in the room shown in Figure 2.2 is *not* a viable option if consistent results are required from freelance staff, because the room could not support the required monitoring accuracy to optimally use the less expensive, less tolerant equipment. Unfortunately, though, such combinations are perhaps encountered in the majority of project studios, which is precisely why we need a further 15 chapters in this book, to deal with the rather more specific issues.

Reference

1 Newell, P. R. 'Not good enough', *Studio Sound*, p. 122 (March 1997)

Chapter 3

Interfacing

Time and time again, when people plan their first studios of any significant complexity, they are shocked once they find the true cost of the cables and connectors needed to connect all the equipment together. They tend to count carefully the cost of the items of equipment when preparing their budgets, but when everything is ready for installation, and the reality of the interconnection costs confront them, there is often not sufficient money left to do the equipment justice. Secondhand leads and connectors are employed, frequently without due thought of the many possible wiring configurations which they could employ. A simple lead, such as one connecting a phono (RCA) plug to stereo jack plug, may seem appropriate to connect the output of one device to the input of another, but there could be several ways of connecting those two 'simple' plugs. Although under many circumstances, a signal will flow between the two pieces of equipment with *any* of the likely wiring arrangements, it is surprising to many people just how 'wrong' some of those connections can be, and how they can degrade the sound.

Figure 3.1 shows a range of possible methods of connecting the two connectors mentioned above. In many cases, these get made in studios for specific purposes, but subsequently get used for other purposes for which they 'look right', but are unsuitably connected. An example would be a cable which was perhaps being used to connect the unbalanced output from a CD player into the balanced tape return input of a mixing console, needing to be typically wired as in Figure 3.1(b) but which had been made to connect the transformer balanced output of a compressor into an unbalanced input of a tape recorder, and which was actually wired as in Figure 3.1(a). Whether or not it works in its new role is entirely dependent upon the type of balancing used in the tape return input circuitry. The result of using such a randomly interconnection cable may be noisy, it may be quiet, it may sound 'perfect' or it may sound degraded. The result of the incorrect wiring may be totally impossible to predict without knowledge of the precise type of circuitry to which it is connected, and which will not be described in most manuals for semi-professional equipment.

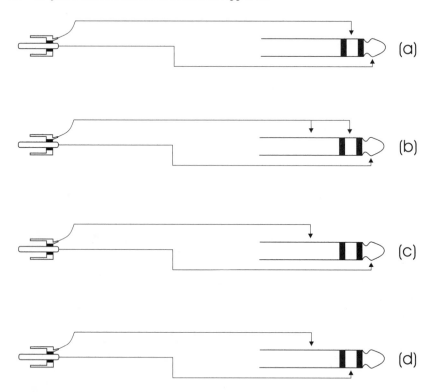

Figure 3.1 Viable RCA (phono) to 3-pole jack configurations. Four possible wiring arrangements which are likely to be found in project studios. (d) is an arrangement which may be found when using some mixing console insert jacks as 'direct' sends, or, in some cases, returns. All of the above have their specific uses

3.1 The origins of the professional interfaces

We currently have a state of affairs whereby equipment manufacturers often produce their equipment with little regard for how compatible it will be in its interconnection with other equipment. This is a more or less inevitable consequence of the evolution of much 'semi-professional' equipment out of the consumer market, where no interconnection standards were ever properly developed. In the realms of truly professional equipment, there is less of a problem, because the professional equipment evolution has been more controlled. In the early days of recording, valve (tube) equipment *demanded* a strict code of interface. Valves could not easily drive long lines directly, so transformer matching was used. The lines were usually run at the +4 dBm (OVU) which was 1.223 volts into a 600 ohm load, a standard which had been adopted by telephone companies as the best compromise between noise and cross-talk in telephone systems. The jackfields (patchbays, in American parlance) were also borrowed from telephone technology, where they were used in the switchboards of telephone exchanges.

Output transformers required accurate terminations if their frequency responses were not to be disturbed, so inputs were terminated with standard 600 ohm impedances. Partly to ease the loading problem of one output feeding multiple inputs (which was a more usual requirement in studios, rather than in telephone systems), the studio industry began to terminate the outputs directly, and not rely upon a 600 ohm loading by the following input. This allowed 10 Kohm 'bridging' (see Glossary) input impedances to be used, and by the mid 1960s almost all professional studio equipment had nominally 600 ohm outputs and 10 Kohm inputs; balanced, of course. Thirty years later, at the top of the professional end, electronic balancing of inputs has become widespread, and the outputs have often tended to become unbalanced (although they are almost universally of very low output impedance; less than 50 ohms). However, there is usually no problem whatsoever in interconnecting any of the truly professional equipment produced over the past 40 years. Old style outputs will drive old or new style inputs, with, at worst, only the addition of a simple resistor, and old style inputs will accept new or old style outputs. About the only complication that arises is that nobody seems to be able to quite make up their mind about which pin should be in-phase [(+) (hot) (live)] on an XLR type connector.

The Cannon XL connector was widely used in the 1950s as a professional audio connector, but it was found that there could be problems of poor contacts after many insertions. Cannon's fix was to use a resilient rubbery material in which to mount the contacts, so that the contacts could be slightly offset, and made to 'wipe' on each insertion or withdrawal. The resilient (R) XL became the XLR connector, and it was soon adopted as a professional recording industry standard connector. In the USA, in the 1950s and 1960s, pin 3 usually seemed to be the accepted 'hot pin', with pin 2 being 'cold' and pin 1 ground (earth). In Europe, pins 2 and 3 were frequently used in the reverse sense. In most cases, the male connector was used for outputs, but in the late 1960s, some Studer tape recorders appeared with male input connectors and female output connectors, probably adhering to the concept that for safety (especially at higher voltages) outputs should not appear on exposed pins. Fortunately, this third 'standard' was rapidly abandoned, certainly for low level signals. Microphones have nearly all adhered to one standard pin configuration, with pin 2 producing a positive voltage in response to a positive pressure at the front of the diaphragm.

The AES (Audio Engineering Society) recommended standard is now for pin 2 hot, but its adoption has not been universal. For some large manufacturers, with so many years of production behind them and so much equipment in the marketplace, to re-configure their 'in-house' standard would (could) lead to great confusion in the interchange of old and new equipment. JBL suffer from this problem with their loudspeaker drive units, on which a positive voltage to the red terminal will produce a backwards (inwards) movement of the cone or diaphragm. The AES standard calls for a positive voltage on the red terminal to produce a positive pressure at the front of the cone, i.e. an outward movement. JBL adhered to a very old standard, which few other companies followed, but produced such a large quantity of material that to change over now would be extremely difficult. There would be no way to inform all their users of the change, so old units would be replaced by incompatible new ones, and total chaos could result, giving JBL products a bad name, solely from the lack of information available to the people making the repairs.

JBL compromised on this problem by making their newer *cabinets* adhere to the standard, so a positive voltage on the red terminal of the box will produce a positive pressure at the front of the loudspeaker, but the drive units remain as they were. Somewhat ridiculously, as JBL have been considered to be 'gods' of the industry by many, I have seen new manufacturers start production with the JBL standard in a belief that if JBL do it, then it must be right. There again, it shows the power of the market leaders. Such is also the case with the old JBL 'slant plates', which they used to use for the acoustic terminations of some of their horns. They were a good theoretical solution to a problem, but they created some unpleasant sonic side effects. Nonetheless, JBL sold so many of these, before withdrawing them from production a full 15 years ago, that they became perceived to be a sort of JBL trademark. Despite JBL now washing their hands of such nonsenses, it is remarkable how many other manufacturers have *begun* production of them, no doubt believing that they 'look professional' because JBL used them.

Anyhow, back to the problems of the professional interconnections. In practice, the pin 2/pin 3 problem is not *too* troublesome. As most reputable microphones use pin 2 as positive (hot), we do not have any polarity problems in that department. Even if we have a piece of equipment which is likely to be connected in different parts of the signal chain at different times, such as a tape recorder or an equaliser, then if it has balanced inputs and outputs, and if it is feeding and being fed from balanced inputs and outputs, it really does not matter which pin is hot. As long as the inputs and outputs are in-phase or, more correctly, in-polarity with each other, the relative polarity of the signal will be maintained. The problems arise when a piece of equipment has XLR connectors on its main inputs and outputs, but jacks, or some other similar connectors, on intermediate inputs and outputs, such as insert points, or compressor side chains. In such cases, the polarity of the inputs or outputs on the jacks will be dependent on whether pin 2 or pin 3 was 'hot' on the XLRs. Operators of mixing consoles need to know the phase relationship between external inputs and outputs and the jackfield, and this usually requires permanent and correctly polarised wiring of the equipment, rather than a lash up of leads which simply look like they will work.

Strictly speaking, I should have been referring to 'polarity' throughout this discussion, and not phase, because a polarity reversal is a 180° phase change at all frequencies. However, the polarity reversal switches on almost all mixing consoles are labelled 'phase', and this terminological misuse has somewhat established itself, at least in the middle order of the industry. Nevertheless, two signals with a relative phase shift of around 15° could be referred to as being substantially in-phase, but not in-polarity. Polarity relates to phase relationships of 0° and 180°, being 'in' and 'out' respectively.

3.2 Jackfields (patchbays)

It would seem to be a fundamental requirement of any studio which dares to charge its clients that the equipment should be wired to a central jackfield, and that this jackfield should be of a three conductor, tip, ring and sleeve type. In first line studios, this is *de rigueur*, but the home studio influence has cast some bad influences on the project studios, and also on some who should know better. The home studios, and many project studios for that matter, have often drawn their experiences, influences and equipment from various sources, including home hi-fi, consumer recording equipment, live (stage) equipment and the professional recording world. This can lead to a very difficult mixture of balanced and unbalanced inputs and outputs, from –20 dB to +4 dBm nominal signal levels, (power level outputs tend to be less costly to produce), and both 2-contact and 3-contact connections. They also tend to use mixtures of equipment which are either double isolated, and hence have no ground (earth) lead, or equipment with standard three-core mains (line/power) cables.

The only effective way to reliably and flexibly interface this rather chaotic mixture is by the use of a good quality jackfield, using 3-pole jacks. There are three basic types of 3-pole jacks, as described in Figure 3.2. Firstly there is the 'stereo' jack, where the plug is like a mono ¼ inch jack with an additional ring connector. This type is used on many headphones. As long as the plugs and the connectors on the jacks are kept clean, they can perform well, and cost considerably less than the more professional types. (Incidentally, in all cases, the 'jack' is the female connector, and the 'jack plug' is male.)

Quarter-inch BPO jacks are preferable, these being again of a ¼ inch sleeve diameter, but the ring and tip of the plug are narrower. They were formerly frequently known as GPO jacks, but the name changed when the General Post Office in the UK became the British Post Office. These are more expensive than the stereo jacks, but come with a variety of very hard, low resistance, and self-cleaning contacts, with the normalling (switching) contacts usually being made from hard metals which will not oxidise, such as gold, platinum, palladium, rhodium, iridium osmium or ruthenium. These are fully professional jackfields, good for 20 years of daily use, but their use can also be a great benefit in project studios where their reliability and lack of signal degradation can save a great deal of time and worry. Switchcraft, the US connector manufacturer, developed a miniature version of these jacks, variously called 'bantam' jacks or 'TT' jacks. TT stands for 'tiny telephone' and bantam is,

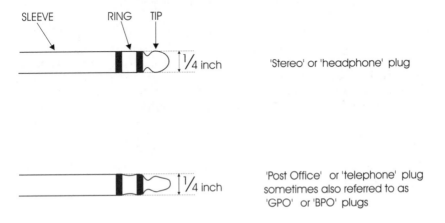

SLEEVE   RING   TIP

¼ inch   'Stereo' or 'headphone' plug

¼ inch   'Post Office' or 'telephone' plug
              sometimes also referred to as
              'GPO' or 'BPO' plugs

Figure 3.2 Two types of ¼ inch 3-pole jack plug. (Note: the 'bantam' or 'TT' jack plug is a ³⁄₁₆ inch version of the GPO/BPO plug)

according to the *Concise Oxford Dictionary*, 'small but spirited', e.g. bantam-weight boxers. Due to their compact size and good performance, the bantam jacks are gradually taking over from the GPO/BPO types, though the plugs can be tricky to wire unless special cable is used. Some of the varieties need special tools for assembly, and may perhaps not be able to be disassembled or re-used. They are certainly nowhere near as robust as their ¼ inch counterparts, which may ultimately be a better choice for project studio use as they are easier to wire, re-usable, and, in most cases, less expensive.

In general, and somewhat perversely, domestic/'semi-professional' equipment is far more problematical to connect together than the very expensive professional equipment. In most cases, a relative novice at recording could successfully connect together the equipment of a fully professional studio, but it usually takes a real professional to wire together a domestic studio, once, that is, it progresses beyond a hard-wired package, and this is where a properly wired jackfield becomes invaluable. Be warned, though; except for the most basic of patching, mono jack, 2-pole jackfields are taboo. They create nightmares in terms of grounding (earthing) which usually can never be correctly resolved for all combinations of cross-patching. Incidentally, the terms 'earthing' and 'grounding' are largely interchangeable, the former being more normal in the UK, and the latter in the USA. However, in these days of space travel, one could take some equipment to Mars and easily ground it, but it would take a devil of a long cable to satisfactorily 'earth' it. One can also 'ground' systems in an aeroplane, but, again, one could hardly 'earth' them whilst in flight. The term 'ground' tends to more specifically refer to a common 'ground plane' to which all voltages in a system are referenced, and it is usually good practice to connect this ground plane to earth, where possible, hence some of the terminological confusion. Section 3.7 and Chapter 4 will explain the situation in greater depth.

3.2.1 Balanced to unbalanced problems

Three-pole jackfields should be wired with the 'hot' (+, in-phase) conductor of a balanced pair, or the signal wire of an unbalanced input or output, connected to the tip. The 'cold' (–, out of phase) conductor of a balanced pair, or the screen of an unbalanced cable, should be connected to the ring. The sleeve (ground) connector should only be used for the screens of balanced cables, and should *never* be connected to any conductor that forms part of an audio circuit, such as the screen of an unbalanced cable. There is a defective logic which is often applied which seems to imply that the difference between a balanced and unbalanced system is that there is no 'neutral', 'cold' connector on an unbalanced system. In fact, they both have a 'hot' connection, and they both have screens, but in an unbalanced system, the screen acts both as a 'cold' *and* a 'screen'. One of the main drawbacks of this is that any interference which the screen shields from the 'hot' wire, will cause a current to flow in the shared cold/screen conductor, so will superimpose itself to some degree on to the audio circuitry. The concept is illustrated in Figure 3.3.

In a 3-pole jack system, therefore, wired as discussed, a balanced to balanced connection will pass through the jackfield as though passing down one continuous cable. An unbalanced to unbalanced connection will do likewise, though without any contact being made with the sleeve (ground) of the jack.

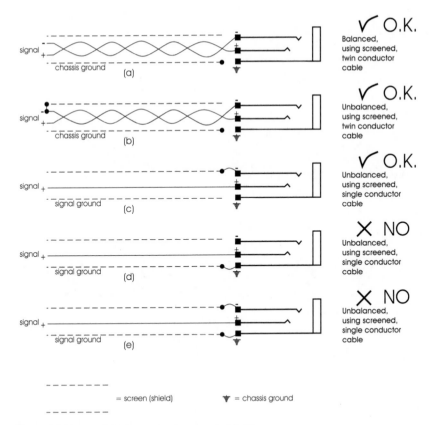

Figure 3.3 Two- and 3-wire connections to a jackfield

A transformer balanced output signal, appearing between the tip and the ring, will automatically connect correctly to the 'hot' and 'cold' connections of an unbalanced input, without any change in the grounding arrangements. In the case of an unbalanced output feeding a balanced input, the situation is entirely dependent upon the type of balanced input being used. Here, there are three basic possibilities, as shown in Figure 3.4. The first is the old, reliable transformer balanced input. This is true balancing, with many megohms of resistance to ground. The drawback to using transformers is that they are very expensive (if they are to do justice to high quality sound equipment) and they usually cannot pass very low frequencies, although 5 Hz or less is possible with some best quality devices. In project studios though, the restricted low frequency response of the more moderately priced transformers is usually of lesser consequence, as the type of equipment chains normally found in such studios will also tend to have limited low frequency (LF) responses. Secondly, and thirdly, there are two types of electronically balanced inputs, of which there are two basic types; one of which *can* accept the 'cold' terminal being shorted to ground, and one which cannot. In the case of the type of balanced input which cannot accept one of its 'legs' being shorted to its own ground, care should be taken, and the manufacturer's advice should be fol-

1) To a transformer input

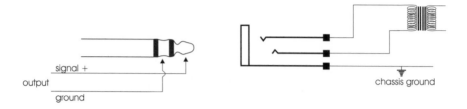

2) To an electronically balanced input

a) with grounded 'cold' (–ve)

b) with floating 'cold' (–ve)

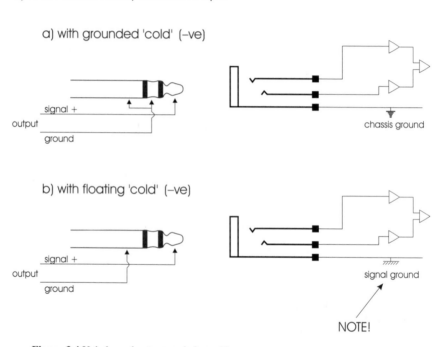

NOTE!

Figure 3.4 Unbalanced outputs to balanced inputs

lowed when connecting such an input to a jackfield. This point is a good example of how, in such situations, a jackfield can be used to standardise all inputs and outputs, so that with only a few exceptions anything can be connected to anything else, in any order. Grounding arrangements can also be optimised, so that hum and other noise pick-up is minimised, though contradictory requirements *can* exist.

Just consider the interconnection, using individual cables, and not a jackfield, in a system which consisted of a Soundtracs Topaz mixing console, a Mackie 1604 as a sub-mixer, a UREI equaliser, and a typical outboard multi-effects processor with balanced jack input and unbalanced jack output. The insert points on the Soundtracs and the Mackie consoles both use a single stereo jack, but whilst the Mackie uses the ring as the send and the tip as the return, exactly the opposite is the case with the Soundtracs inserts. The UREI equaliser has XLR-type inputs and outputs, so if a lead was made to insert this equaliser in the Soundtracs console inserts, it could not be used with the Mackie inserts, because the console *send* would be connected to the female XLR, which connects to the UREI *output*. If a multi-effects processor were to be used having stereo jack inputs *and* outputs, obviously, the plugs could be reversed, and a single lead could be used for connecting the processor into either mixing console. Perhaps! Or there again, perhaps not.

There are numerous pitfalls which can be run into here. For some equipment, especially when the equipment is grounded (as it should be) via its mains lead, there may be a preferential requirement to lift the screen of either the input or the output in order to achieve the best noise rejection. Obviously, such a regime could not be followed with reversible leads. However, we would have even bigger problems if the equipment to be inserted had, as is commonly the case, electronically balanced inputs, and unbalanced outputs. To the inexperienced operators, there would be no apparent reason why reversible leads could not be used, and indeed they would probably work in many situations. This false sense of security could lead to great time wasting, though, when such 'tested' leads were the actual source of problems in another application for which their grounding configuration was inappropriate. Unfortunately they may be the *last* thing to be suspected, hence the time wasting.

3.3 Jacks – 2- or 3-pole?

An absolutely ludicrous state of affairs exists in terms of the use of the mixture of 2-pole and 3-pole jacks. It is frequently the case that much equipment using ¼ inch jacks gives no information whatsoever as to whether the inputs and outputs are unbalanced or balanced, and whether or not they use 2-pole or 3-pole jacks. Even many instruction manuals give scant information about the terminations. Looking down into the jacks with the aid of a penlight is one way to find out if it is definitely unbalanced, which it *must* be if there are only two contacts on the jack. Some equipment, however, uses 3-pole jacks as standard, so even though it appears to be balanced, it may still be wired unbalanced. Even removing the top and bottom panels from the equipment may not readily reveal the answer, as double-sided boards are often employed and the jacks may cover the tracks. The difficulty of discerning the precise nature of the jacks inevitably leads many people to a 'try it and see' approach.

In fact, there would appear to be no reason whatsoever, except for the disgraceful excuse of saving an outrageously small amount of money, why, 2-pole ('mono') jacks should *ever* be used on equipment intended for studio use, or even on musical instruments having line level connections and which are likely to be used in studios. I suppose that all of this goes back to the electric guitar and amplifier, with the historical, standard use of mono jack plugs and high impedance leads. Electronic keyboard instruments tended to follow the same code, but the evolution of devices for use in conjunction with them conflicts badly with the professional studio wiring approaches. If only *all* new equipment used 3-pole jacks, wired in accordance with that suggested earlier for jackfields, then it would still be mono jack plug compatible, as the mono,

2-pole plug would always short the ring and sleeve together, hence, automatically re-making any separated contacts which needed connecting for unbalanced operation. This would preclude the use of certain types of electronic output balancing arrangements, though (the ones which will *not* accept the cold [–]ve being connected to ground), but as these can often create more problems than they solve, their demise would be no great loss. However, as that is not the case, let us return to our attempts to insert the processor into the system which we were discussing a few paragraphs previously.

3.4 Hidden problems in tolerant systems

If we were to make up a lead for use with the Mackie console, then the ring (output) of the stereo insert jack would connect to the tip of the stereo, device-input jack. It would be wise to use microphone type (twin conductor and screen) cable, so this would be connected as standard at the device input jack, and with the 'cold' connected to the screen of the insert jack, the screen not being connected here. Typically, the processor would use a mono output jack, and if 'destination only' screening were to be adopted, then the tip of the output jack plug would connect to the tip (return) of the input jack. The cold connector would be connected to the body of the mono jack plug, and the cold and screen would be twisted together and connected to the sleeve of the insert jack. The arrangement is shown in Figure 3.5. If it should then be attempted to use this lead with the Soundtracs console, it would be necessary to use the input connector for the output, and the output connector for the input. Obviously, the 'destination only' grounding would become 'source only' which is another option, and may or may not be acceptable depending on the characteristics of the equipment in use. The mono jack plug would short the ring to the sleeve on the input, unbalancing it. Again, dependent upon the nature of the electronically balanced input on the processor, this could either work normally, or signal degradation could result. The output connector would connect its tip (hot) correctly, but the 'cold' would be left floating, as the 2-pole output jack would have no ring connector. The ground connector of the output jack would connect to the screen of the stereo jack plug, but this was left disconnected at the insert jack plug, to avoid ground loops, so the only way that the output ground could reach the insert would be via its connection to the input jack screen, the short circuit which the mono jack created between the input 'cold' and ground, and then via the input lead 'cold' cable, and back to the insert ground. The input cable 'cold' in this case would carry the send *and* return signals, with both being rendered unbalanced. This could risk instability, radio frequency interference, and harshness of the sound, not to mention the signal degradation which could result from shorting the input 'cold' to ground. Yet, entirely unwittingly, these nonsenses are perpetrated daily, around the world.

If the above paragraphs have passed over the heads of some (many?) readers, then they will have still served a purpose. They will have made it crystal clear that the seemingly simplest of interconnecting operations can even be mind-boggling for experts. The super-irony lies in the fact that these interfacing problems largely only exist amongst equipment intended *specifically* for use by non-experts; whereas the

MACKIE INSERT CONFIGURATION

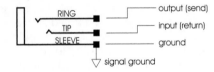

SOUNDTRACS INSERT CONFIGURATION

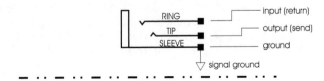

(a) Insert lead for connecting an effects unit, having a balanced input and unbalanced output, to the insert jack of a Mackie console

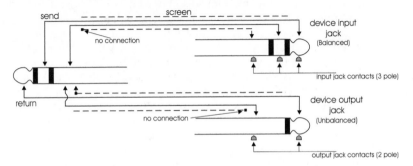

(b) The same lead, used in reverse with a Soundtracs console

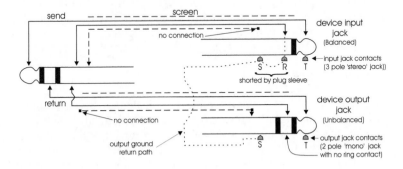

Figure 3.5 Cable confusion

equipment designed for use by experts is usually completely free of any such conundrums.

The non-standardisation of connectors creates these absurdities, and it is only reasonable to expect that, during the pressure of the session, a re-patch will be made with the first lead which comes to hand and which 'seems' to work. The problem is insidious, because the way in which hi-fi turns to low-fi is not always as apparent as it is with professional equipment, where an obviously wrong termination is likely to lead to a thin 'one-legged' sound with significant loss of level, and/or distortion. Often, with less expensive equipment, there is no such obvious restriction in the frequency response, no obvious distortion and no excessive noise. The signal can simply just lose its sparkle, its openness, its naturalness and, in stereo, its imaging. Obviously, if this sort of arrangement is used with a piece of unfamiliar equipment, or an unusual combination of equipment, there may be no prior knowledge of how good the sound *could* be, so the degraded sound, due to poor interfacing, is either accepted as it is, or the equipment is maligned as being poor-sounding.

If we were now to try to use the Soundtracs and Mackie mixing consoles with an expensive, professional Lexicon reverberation unit, a different set of problems may arise. Some of the more expensive Lexicon devices have excellent input and output circuits which tolerate just about any connections other than a direct short between pins 2 and 3. Lexicon therefore do not declare a preference of pins 2 or 3 for hot or cold. In many cases, mixing consoles of the lower and middle price ranges may well use different connectors on sends and returns. It is quite likely that a send would leave from a jack, used for auxiliary outputs, and return into a balanced input, which may well be an XLR-type connector. If loose, separate cables were used to connect such equipment, then just about any arrangement of mono or stereo jack plugs, connected to XLRs with either pins 2 or 3 wired as hot, and even with either pins 2 or 3 left disconnected, would allow the system to operate. Unless the wiring of each cable was very carefully verified, and it was known that the wiring standards were consistent, it would be very possible to unwittingly use the Lexicon without ever realising that anything was wrong. Quite possibly, the inconsistency between input and output polarity, and between that of left and right, also, would not allow the sound of the programmes to be heard at their optimum, or as intended, but, as there is little phase coherence in reverberant signals, it may again be unapparent that optimal results would *not* be being obtained. In fact there may be nothing 'wrong' with the sound per se, but it may well not be as intended by the manufacturer.

3.5 The chaos of discrete leads

For rapid and reliable work, there is therefore no real alternative to using a 3-pole jackfield and connecting to it each piece of equipment, after having very carefully checked the wiring standard of each connector, either physically, electrically, or via the manual. By this means, consistent grounding systems can also be achieved, because the options of signal grounding can be very different depending upon whether or not a particular piece of equipment has a 2-core, or 3-core mains lead. All of these things can duly be taken into account when a permanent installation is arranged, and provisions can easily be made for interfacing extra equipment brought in from outside. This can be done by means of cables with a range of connectors and wiring systems on one end, but all having suitable standardised plugs for whatever type of jackfield is being used on the other end. It is, of course, imperative that the leads are clearly labelled, showing exactly how they are configured.

It is true that jackfields, plus 40 or 50 patch-cords, are not cheap to buy, but nor would it be cheap to buy all the necessary permutations of other connectors and leads which would be necessary to be able to make the optimum interconnections between *all* likely combinations of loose equipment. If errors were to be avoided when concentration was being applied elsewhere (such as on the musical aspects of the recording), then each piece of equipment would need to be clearly labelled as to how it should be wired. Furthermore, each *lead* would need to be clearly labelled at each end, explaining the pin connections, and a large stock of leads would be needed to allow for all likely combinations of equipment. The biggest problem would come in remembering the optimal grounding system to be used with each set-up, and consequently whether the leads should be ones with the shields connected at both ends, or just the input or the output end. In all cases, only balanced-type cable, with a good shield, should be used, and never guitar-lead type cable, with just a single core and screen, as there would be no way, in this case, of keeping the signal −ve separate from the screen.

There is no realistic alternative to a good quality 3-pole jackfield if a professional service is being offered, or if good quality sound is to be expected as normal, as it would be unreasonable to expect even an experienced technical person to remember all that was required, especially if work was rapid and/or pressurised. Clearly, to do things properly with a discrete lead system is impracticable, though inexperienced people do use such systems. It is really outrageous how so many mixing console manufacturers produce 30- or 40-channel consoles entirely terminated by individual connectors on the rear panels. This is an unashamed marketing exercise, as they are producing consoles which, in almost any serious operation, need to be connected to an appropriate jackfield. Their argument is often that if the greater part of the customers for such consoles are too ignorant to realise what is truly necessary, then if the other console manufacturers are selling 'stripped down' versions, they would be unable to compete (and hence to survive) if they produced more expensive products. They do have a valid point, but the real nonsense is that it probably costs the customers twice as much (at least) to fit an external jackfield than it would cost to fit one at the factory, during production.

When recently reading Ben Duncan's book *High Performance Audio Power Amplifiers* (Newnes), I came across the following quotation which is so apt for this situation. It was from John Ruskin (1819–1900): 'There is hardly anything in the world that some man can't make just a little worse and sell just a little cheaper, and the people who buy on price alone are this man's lawful prey.'

3.6 Multiple signal path considerations

What should also always be remembered is that a studio system is not just a big hi-fi system. A hi-fi system usually consists of a pair of signal channels, with serial connections of no more than about three active devices. Typically this would be a source (CD, tape, radio etc.), a pre-amplifier and a power amplifier, there are usually no auxiliary send and returns which are in simultaneous use with the main signal channels. There are therefore no parallel signal paths, except the left/right pair itself. Despite this apparent simplicity, many hi-fi enthusiasts go to great lengths to ensure the good quality of inter-connect cables and plugs. The attention which they pay to this aspect of the system is often much greater than most studios pay to the inter-connection of vastly more complex systems, and this fact should help to put the problem into perspective. I realise that there are many hi-fi 'freaks' who go well over the top, and I am not advocating spending $2000 on each pair of phono (RCA) leads, but some of the real rubbish that does get used in ostensibly pro-fessional studios is inexcusable. In fact, one reason why hi-fi enthusiasts pay vast amounts for interconnect cables is that much hi-fi equipment does not have output stages which can adequately drive the poor quality interconnect cables. Top quality studio equipment tends to have output stages of much lower impedance and higher output capability, which are less prone to the slew rate limitations that can occur with much domestic hi-fi equipment. Good quality studio equipment is *designed* to drive long interconnect cables of standard quality. Hence, the benefit from using esoteric hi-fi interconnects may not be relevant. Good quality standard cable will usually suffice.

Once we leave the realms of the 'simple' serial signal paths, and enter the realms of multiple channels and parallel signal paths, a completely different approach needs to be taken to the system interfacing. A studio set-up is much more complicated than a domestic hi-fi system, and the opportunities for interference from mains-borne noise, radio frequency interference, and inter-ference induced from digital equipment are enormously greater. Screened pair (balanced) type cables are virtually mandatory if noise and interference are to be kept out of the system. Remember, never should any studio cable have signal current flowing down a screen. Furthermore, screens should ideally be connected to ground at one end only. The only exception to this is microphone cables, where the microphone itself is generally not connected to a mains earth. The wiring practices for hi-fi and studio systems are thus not necessarily directly applicable to each other.

3.7 Grounding of signal screens

Rather disgracefully, the equipment manufacturers do a lot of nonsenses themselves, and much of this, it must be said, is clearly out of the ignorance of equipment designers themselves. Basically, for good interfacing, all equip-ment should use 3-contact connectors, with the screen connection made nowhere else other than the chassis. Neve were doing this with their mixing consoles in the 1960s, and, even then, the practice was based on 20-year-old literature, so it seems that there are some slow learners around. Neil Muncy outlined what he called 'The pin 1 problem' in a classic AES paper,[1] in which he clearly threw down the gauntlet to manufacturers.

In another paper in the same journal,[2] Stephen Macatee very concisely reinforced the same point, and on some aspects of bad interfacing went even further. From that paper, the following quotation is taken:

The 'Pin 1 Problem'
Many audio manufacturers, consciously or unconsciously, connect bal-anced shields [screens] to audio signal ground – pin 1 for three-pin (XLR-type) connectors; the sleeve on ¼ inch (6.35 mm) jacks. Any cur-

rents induced into the shield modulate the signal reference to that ground. Normally, great pains are taken by circuit designers to ensure 'clean and quiet' audio signal grounds. It is surprising that the practice of draining noisy shield currents to audio signal ground is so widespread. Amazingly enough, acceptable performance in some systems is achievable, further providing confidence for the manufacturers to continue this improper practice – unfortunately for the unwitting user. The hum and buzz problems inherent in the balanced system with signal grounded shields have given balanced equipment a bad reputation. This has created great confusion and apprehension among users, system designers as well as equipment designers.

Similar to the 'pin 2 is hot' issue, manufacturers have created the need for users to solve this design inconsistency. Until manufacturers provide a proper form of interconnect uniformity, users will have to continue their struggle for hum-free systems, incorporating previously unthinkable practices.

The 'unthinkable' practices, to which Stephen Macatee refers, no doubt include the 'lifting' or removal of safety grounds on equipment for which they are an essential part of the design. Experience has shown, though, that there is an enormously greater number of studios who violate this legal and sensible requirement than those who completely comply with it. The problem is that with many equipment combinations, even of very well-known equipment, there can simply be no other way to achieve a hum-free system. Well, it *may* be possible by fitting correctly configured input and output stages, or by adding separate line termination amplifiers, but this could double the cost of the installation. The reality is that, human nature and economics being what they are, if the cost is going to significantly increase, then the ground comes off; dangerous or illegal as it may be. There is absolutely no doubt where the blame lies for this situation; it is rooted in the ignorance and greediness of equipment manufacturers, together with the users who demand ever-cheaper equipment.

3.8 Balanced versus unbalanced – no obvious choice

Balanced systems have two inherent advantages over unbalanced systems: the ability to reject more noise and interference, and the ability to completely free the signals from a noisy ground. Unfortunately good balancing arrangements tend to be expensive, and this does not fit in well with the cost conscious philosophies of most of the more competitively priced recording studios. There are situations where nominally balanced inputs are not very well balanced at all, and their performance can be problematical. There are other situations where the *un*balanced inputs are preferable on equipment which has dual inputs, simply because it is easier to make a good unbalanced input than a good balanced one. Frequently, it seems that balanced inputs of dubious quality are installed merely to make equipment *look* more professional, and hence they are probably no more than a marketing ploy. In many cases, unless there are great induced noise problems, the use of a good unbalanced input will be preferable to the use of a poor balanced one.

Good balancing requires the use of multiple chips, discrete components, or high quality transformers. If an inexpensive piece of equipment sports a balanced input using a single chip, then its quality of performance is to be questioned, and it should *not* be used as an automatic preference. Ben Duncan has written much about this subject, and it is worth consulting much of this work.[3] Furthermore, balanced to unbalanced terminations, in almost all cases except for high quality, expensive, transformer balanced, floated inputs or outputs, do not make happy partnerships. Bearing this fact in mind, it obviously becomes difficult to generalise about which input to a piece of equipment with dual inputs is the 'best'. It may be that the balanced input is sonically the best one to use with balanced outputs, but the unbalanced input should be used with unbalanced outputs. It is difficult to give absolute advice here. The answer may depend on cable length, or other factors.

The toughest problem to solve is usually the connection of certain types of electronically balanced outputs to unbalanced inputs. In fact, to avoid the possibility of mis-termination when some of the outputs may be connected, from time to time, to various different inputs, it may be wiser to find the optimum way to unbalance the outputs, then leave them that way. To do this, electronically non-floating outputs should leave the out-of-polarity (−ve, cold) connection disconnected. Electronically pseudo-floating outputs should have the out-of-polarity connection shorted to ground at the output terminals. It is important to ground it at the output, because if it is remotely grounded, the low output impedance of the device can sometimes drive ground currents through any wiring loops, and can introduce distortion into the in-polarity side of the system.

Again, as mentioned in Section 3.4, in conjunction with the use of non-standard leads, equipment wrongly connected in this way can often operate inadequately but with the problem going unnoticed, because unless the clean sound has been heard, the degraded sound may well be taken as '*the* sound', and equipment may gain a poor reputation due solely to incorrect termination. I will not go so far as to say 'gain an *unjustified* poor reputation' because the reputation may well be justified by the fact that the manufacturers have chosen an output topology which is prone to this sort of problem in its likely theatres of use. Having balancing transformers available on the jackfield to interface incompatible inputs and outputs when patching in effects is a very useful facility, though good transformers are essential. In fact this is sometimes the *only* practical answer to the problem.

3.8.1 Inverted logic?

One absolute nonsense which I have encountered on a number of occasions is when a relatively inexpensive piece of equipment is fitted with such connection-sensitive outputs, and when the manufacturers must know, full well, that mis-terminations are as likely to be the rule as the exception. It is typical to find that the higher-end equipment, which is expected to be in the hands of professionals, will be fitted with less finicky outputs, whilst the equipment destined for distinctly non-professional use is supplied with outputs which can even catch the professionals off-guard and cause problems. I know this, because I have *personally* been caught out by such equipment. The audio recording equipment industry *desperately* needs to get its collective act together and straighten out these issues, because there is absolutely no hope of educating the market about all the correct termination protocols.

3.8.2 Sixteen options for one cable

In Stephen Macatee's AES paper[2] he listed sixteen possibilities for connecting different combinations of balanced and unbalanced inputs and outputs, the number being compounded by the possibilities of the screens (shields) being connected to signal ground, or chassis ground. Of the sixteen possibilities only four could be optimally made with 'off-the-shelf' cables, and even one of those had exceptions. Of the other twelve, one to one connection of the three contacts of the plugs on each end of the cable could not be made. This reinforces the earlier statement about the need for a permanent jackfield, as the requirement for specifically connected leads is sure to end up in disaster if standard, or even a selection of non-standard cables are used. None of this, incidentally, is referring to pin 2 or pin 3 hot anomalies. These sixteen possibilities presumed a standard pin 2 hot. Figure 3.6 depicts the different possibilities for interconnections with signal or chassis grounds.

The other great problem with 'lash-up' connections is that even when a system is made to work noise-free by ground lifting on the mains leads or the signal cables, it only needs the substitution of one piece of equipment anywhere in the system, even remote from the direct signal chain, to upset the grounding. In such an instance it may that it would be necessary to begin again, from square one, with the whole problem of noise solving. Even the

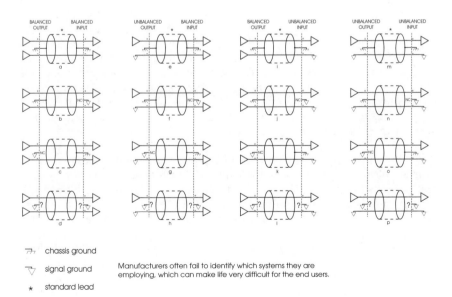

chassis ground

signal ground   Manufacturers often fail to identify which systems they are
                 employing, which can make life very difficult for the end users.
★   standard lead

Figure 3.6 Sixteen different optional XLR interconnections. Each are the optimised connections for the indicated situations. Note that only four, at the most, can be made with standard 1 : 1 leads (after Macatee)

concepts of 'source only' or 'destination only' screen grounding, each of which have their disciples, can be very system-specific in terms of which is best. The truly unfortunate thing is that all of these problems are loaded on to the owners and operators of the equipment due to the lack of determination from the equipment manufacturers to conform to unified standards of wiring. It is actually a disgrace and shame upon the manufacturers who perpetrate much of this nonsense.

3.9 Some comments

The number of possible, or even *likely*, equipment set-ups in project studios is enormous, and each one will present its own specific set of problems which needs to be solved in order to achieve a satisfactorily noise-free system. No simple set of rules can therefore be offered about how the problems can definitely be solved for each individual case, but at least an understanding of the possibilities can act as a guide. What is more, an awareness of the nonsenses which have led to this state of affairs existing will reduce the frustration which may be felt when some systems stubbornly refuse to perform without the attendant hums and buzzes. An understanding will also help to ensure that time spent troubleshooting is used in a meaningful way, and is not spent going round in circles.

Another aspect of a well-interconnected system is how relatively insensitive it is to the grounding (earthing) arrangement used for the power (mains) wiring. Good earthing is mandatory for the safety of the system, and 'clean' mains is always preferable to 'dirty' (noisy) mains. However, special grounding arrangements such as 'star' earthing are much less of a requirement for noise-free performance when the audio interconnections conform with the preferred practices. The subject is huge, and is dealt with in great detail in the book by Philip Giddings, referred to in the bibliography at the end of this chapter. Five hundred and fifty-one pages on audio systems wiring! The book is absolutely mandatory reading for anybody who is seriously studying the problems of audio wiring, but Giddings also calls for a more responsible attitude from the manufacturers in terms of making equipment to agreed standards, and lifting the burdens off the customers who provide their income. Manufacturers are receiving a lot of bad press on these points, and deservedly so.

The optimisation of sonic purity in many project studios is beyond the capacity of many owners, operators and installers of such facilities. This is a reality of the situation in which we find ourselves. Hopefully, though this chapter may not only give some guidance as to how to circumvent many of the common problems, but will also serve to relive some of the frustration which people may feel when a system which is thought to be relatively simple obstinately refuses to behave correctly. One point to bear in mind at all times is that, in terms of their interconnections, many project studios present vastly more complex problems than ultra-expensive studios which use the finest equipment available. When a lesser system fails to operate to perfection, try not to put all of the blame on the technical personnel who installed it, because they may have inherited many intractable problems which have originated in the sloppy practices of many equipment manufacturers. What is more, if you are the owner or operator of a semi-professional studio, who has to live with these problems daily, do not feel too inadequate if you cannot readily find fixes. You have probably had the problems dumped on you by the manufacturers.

It is rather absurd that as a result of this, the most inexpensive equipment requires the most expensive installation, but that is what market forces have led to. Glossy advertisement, big claims, low price, sell it to the ignorant, then run like crazy before they find that it will not interface very well because all the corners have been cut to keep the price low and pay for the expensive advertising. It is a disgraceful state of affairs.

The fact that this chapter has been rather long-dwelling on the problems and somewhat short on answers is indicative of the current realities. Chaotic free-for-alls are not conducive to good engineering solutions, but for the moment, we simply have to live with this rather unpleasant reality.

References

1 Muncy, Neil A., 'Noise Susceptibility in Analog and Digital Signal Processing Systems', *Journal of the Audio Engineering Society*, Vol. 43, No. 6, pp. 435–53 (June 1995)
2 Macatee, Stephen R., 'Considerations in Grounding and Shielding Audio Devices', *Journal of the Audio Engineering Society*, Vol. 43, No. 6, pp. 472–83 (June 1995)
3 Duncan, Ben, 'The New Age of Radio Defence', *Studio Sound*, Vol. 38, No. 10, pp. 85–8 (October 1996)

Bibliography

Davis, D. and Davis, C., *Sound System Engineering*, Focal Press, Boston, USA (1997)
Giddings, P., *Audio System Design and Installation*, Focal Press, Boston, USA and Oxford, UK (1995)
Giddings, P., *Audio System Design and Installation,* Howard W. Sams, Indianapolis, USA (1990)
Journal of the Audio Engineering Society, Vol. 43, No. 6 (June 1995) issue dedicated to papers on audio system grounding and interconnecting.

Chapter 4

Mains supplies and earthing systems

Large professional studios usually have 'technical earth' systems, where a dedicated earth is sunk into the ground, to which no equipment is connected other than the audio equipment. Sometimes this earth is isolated from other earths, and sometimes it is bonded to the earth which is used by the electricity supply company. The regulations about this vary greatly from country to country, and even from region to region, so in a book such as this, designed for an international readership, it is only possible to discuss things in general terms. However, wherever you are, and whatever path you seek to follow on this subject, it is imperative to discuss matters with a local, licensed/approved electrician, who is fully aquainted with the local regulations. There is an inherent problem in this, though, because many qualified electricians are totally unfamiliar with the concepts of technical earths, and even professional studios sometimes encounter problems with such negotiations.

Nonetheless, the object of the exercise is to provide a ground reference plane of a low impedance to earth, which acts as an enormous 'sink' for all electrical 'dirt'. As the impedance of the surface of the planet is so exceptionally low, and its mass is so great, if one can get as low impedance a path as possible to earth, then whether the technical earth is, or is not, connected to an electricity company safety earth is usually not of too much relevance. The task is to get the audio system earth connected to any other earths in such a way that any common impedances are minimal. Figure 4.1(a) shows two different, right and wrong approaches which should be more or less self-explanatory. Any readers unfamiliar with the concept of impedance, or the benefits of low impedance sources, should refer to the glossary at the end of the book.

Geography, or rather local geology, plays a great part in the effectiveness of earthing systems, and if a studio is sited in a region of poor ground conductivity then little can usually be done about it. However, as mentioned in the last chapter, when the audio system interfacing is carefully and correctly terminated, then the necessity for an excellent earth is significantly reduced. On the other hand, when the earthing system is excellent, an audio system may stand a better chance of tolerating less than optimum audio interfacing without the manifestation of too many problems. Obviously, though, one should strive to get both aspects as good as one can.

The general level of electromagnetic interference has risen, gradually, in recent years, to a point where we now seem to be swamped with it. The

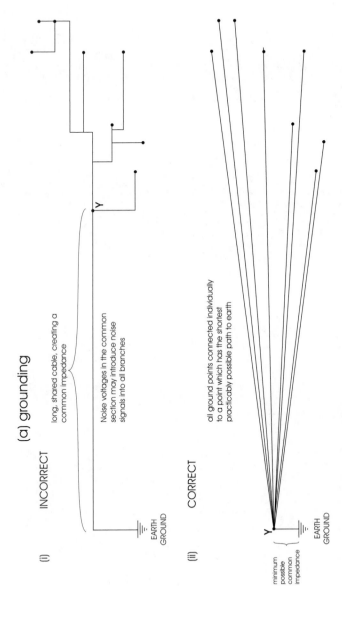

Figure 4.1 (a, b) Grounding and power distribution: (a) grounding

(a) grounding

INCORRECT

(i)

long, shared cable, creating a common impedance

Noise voltages in the common section may introduce noise signals into all branches

EARTH GROUND

CORRECT

(ii)

all ground points connected individually to a point which has the shortest practicably possible path to earth

Y

minimum possible common impedance

EARTH GROUND

The section of the cable between point Y and earth, in each case forms the lower half of a potential divider between the individual earthing points and earth ground.   The impedance of this section should be minimised, which means that the series components of resistance *and* inductance should be carefully considered. As we are dealing with AC, it is futile using a path of minimum resistance if the inductance is still allowed to remain high. In the case of power wiring, the close coupling of the individual cables for live and neutral can be used to help to cancel the inductance in each strand, but with single earth (ground) cables, this option is not available, so short runs are the only real option.

(b) power distribution.

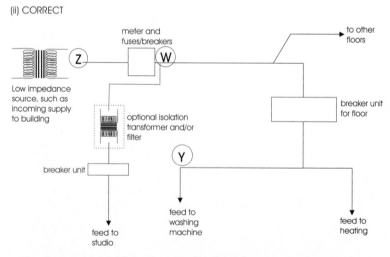

In the above arrangement, the long, shared section of the supply, between points Y and Z serve to increase the impedance of the supply to the studio. Electrical loads on the other floors, together with current taken by the heating and other devices, will cause the voltage at Y to fluctuate as the current demand varies. Furthermore, interference caused by the washing machine motor will be superimposed on the supply line to the studio. The result would be a dirty and unstable supply to the studio when other electrical systems in the building were in use.

Here, the studio is fed directly from the meter at the incoming supply to the building, via over-sized cable. Whilst the voltage at point Y may still vary as other equipment in the building is turned on and off, the studio feed, from point W, will take advantage of the stability of the low impedance supply and will remain relatively stable. The section Z-W, which is of only minimum length and impedance, will form the lower half of a potential divider, Y-W-Z, with most of any interference from washing machine motors, or similar devices, being dissipated in the upper portion, Y-W. This solution may be seen as 'inconvenient' by the electrical installer, and unnecessarily expensive from a conventional power wiring point of view, but it is the only way to ensure an adequate supply to sensitive recording equipment.

Figure 4.1 *cont.* (b) Power distribution

amount of radio traffic is now enormous, which has polluted the air with electromagnetic signals of all sorts. Even in the earth, surprisingly large noise currents can flow near to railway lines where digital signalling and control
equipment is used. In industrial areas, there can also now be enormous
amounts of electromagnetic pollution in the air, the ground and the mains
electricity supply. Staying clear of all this is no mean feat, but by paying due
attention to each potential source, the effects can usually be reduced to
insignificant levels. Unfortunately, the types of equipment often found in
project studios tend to be more prone to external interference than top-line
equipment. This problem is compounded by the fact that the level of knowledge about how to deal with the noises is less readily found in project studios.
I will try to simplify this chapter as far as I can, but as the problems are not
simple, then neither will be all of the solutions.

Good audio interfacing practices and good earthing practices can go a long
way to reducing interference problems, but they may be able to do little about
mains-borne interference. The European Union and other authorities around
the world have already introduced legislation which restricts both the amount
of interference which equipment can generate, and its sensitivity to it, but this
will take years to take real effect. In fact, due to the seemingly endless growth
of the use of electrical and electronic equipment, much of the legislation may
only serve to slow down the rate of growth of electro-magnetic pollution,
rather than actually reducing it from present levels. The two cardinal rules are
to ensure that studio equipment is supplied with the lowest practicable source
impedance for its electricity supply, and to keep all the interconnected equipment on the same phase. Figure 4.1(b) illustrates the right and wrong ways of
installing power wiring if interference problems are to be avoided.

4.1 Low impedance supplies

Many pieces of equipment used in studios, such as most power amplifiers, are
notorious for drawing current in surges. Figure 4.2 shows the typical way that
an amplifier (other than the inefficient, constant output-stage current, class A
designs) will draw a large transient current from the mains in response to a
loud bass drum signal being driven into the loudspeakers. The amplifier will
initially draw current from the reservoir capacitors in its power supply; but,
once the voltage rails fall below the level where the peak voltage on the input
rectifier exceeds it, large currents may begin to flow to replenish the reser-
voirs. The power drain from the electricity supply is thus not a continuous
affair, but a continual series of high current pulses. As the supply impedance
is in series with the amplifier, a potential divider will be formed, with the
amplifier forming the lower element. When the amplifier calls for a current
surge, its mains input impedance will drop, and unless the power source is of
a significantly lower impedance, the mains supply voltage, measured at the
wall socket, will decrease as each pulse is drawn. This will lead to a voltage
waveform as shown dotted in Figure 4.2, and this waveform represents a
supply containing much harmonic distortion.

It matters naught whether or not there is a power conditioner on the mains
input to the studio, because these harmonics, which can extend up to rather high
frequencies, are being generated locally, and so they will have the same effect

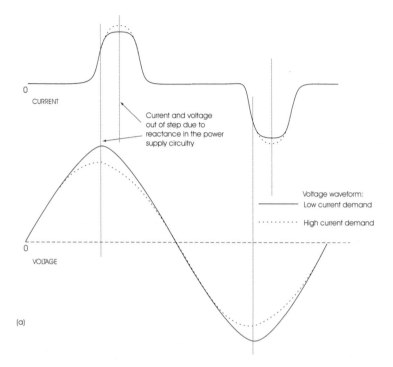

(a)

In the case of a stable, resistive load, such as an electric heater, the current demand would follow the sinusoidal waveform of the voltage, and both would be in phase. An audio power amplifier, however, presents a very different load to the supply. Due to the presence of inductance and capacitance in its power supply circuitry, it will present a partially reactive load to the supply, in which the current and voltage peaks will be displaced in phase. What is more, the current demand will not be sinusoidal, but will be more like the waveform shown above.

As the output stages of the amplifier supply power to the loudspeakers, they will deplete the charge in the reservoir capacitors. In a conventional power supply, the capacitors will be replenished via a transformer and bridge rectifier, but the bridge cannot begin to conduct until the voltage on the AC side exceeds the DC voltage on the capacitors. This means that as the AC voltage rises from zero, no current will be drawn until the voltage on the transformer secondary winding exceeds the residual voltage in the capacitors. The above figure shows no current initially being drawn, followed by a sudden inrush of current once the DC voltage level is matched. The current then ceases, relatively abruptly, once the AC input voltage of the rectifier falls below the charge voltage on the capacitors. Note that as the current demand increases, the voltage falls, as shown by the dotted lines. The amount of voltage drop is dependent upon the supply impedance. A supply of zero impedance would suffer no voltage drop, which is why very low impedance supplies are often referred to as constant voltage sources.

The dotted lines in the current plot show the likely demand from a zero impedance supply, but the impedance of a poor supply would limit the current which can be drawn. This would cause the voltage to sag, and could cause the voltage on the studio power outlets to look something like the dotted lines on the voltage plot. This, in turn, produces harmonics on the supply voltage, which can extend into the hundreds of kilohertz region. Such harmonics can easily enter recording equipment, creating harsh sounds, and can play havoc with computer systems.

The other thing to bear in mind is that when current is drawn in this way, it is not as would be calculated by a simple V x I calculation. For example, a heater consuming 1kW from a 230 volt supply would draw 4.35 amperes. However if current was only being drawn 25% of the time, as could be the case with an amplifier, then 1kW (or rather 1kVA) would draw 17.4 amperes. From the point of view of an electrician, a 5 amp cable would suffice, because the time-averaged heating effect in both of the above cases is the same. The need for oversized cabling in studios is thus not due to cable heating or power consumption, but to avoid the volts drop associated with the high current pulses, as these cause the voltage waveform distortion which gives rise to the intrusive harmonics.

Figure 4.2 (a, b) Current and voltage interaction. (a) Current demand cycle of a typical power amplifier under heavy drive conditions

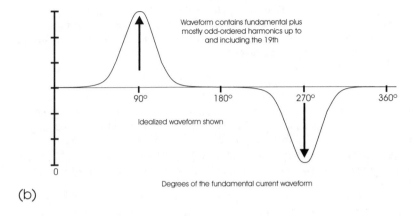

Waveform contains fundamental plus
mostly odd-ordered harmonics up to
and including the 19th

90°   180°   270°   360°

Idealized waveform shown

Degrees of the fundamental current waveform

(b)

Figure 4.2 *cont.* (b) Typical high-peak current waveform for switch-mode power supply input on mains power circuit (after W.H. Lewis)

as any externally generated harmonics entering via a non-conditioned mains electricity supply. Harmonics on the supply can lead to a whole raft of problems, which will be discussed as this chapter progresses. However, if the supply impedance is low enough, it will tend towards acting as a constant voltage source, and within reasonable limits will continue to supply a clean waveform even when high current demands are made. This is exactly akin to an amplifier with a high damping factor acting as a constant voltage source to the varying load which is characteristic of most loudspeakers. Sonically, this means more punch in the bass response, but unless the amplifier itself draws current from a low impedance supply, the punch may be lost if its own supply is depleted by the transient signals, especially when working near to its power output limits.

The power supplies of many pieces of equipment are not very good at filtering out higher harmonics on the mains. In all transformers there exist leakage capacitances between the windings, and these can act as easy routes for the higher frequency harmonics to enter the signal circuits. In analogue audio equipment they can cause harshness in the sound, but, somewhat surprisingly to many people, in digital equipment they can cause glitches and even crashes. They can also introduce errors into control circuits, and jitter into digital audio signals. It is remarkable how many such problems can be solved by using a low impedance supply.

Wherever possible, a studio should draw its mains supply directly from the electricity company's input to the building, and not from any branch circuits. It should not share any cable run with any other part of the building, and the audio supply should not even share any cable runs with the studio's own air conditioning units or refrigerators. It is also worth using the heaviest sensible gauge of wire, even if the fusing is much below this rating. This one can sometimes be difficult to get the electricians to understand, as, to many of them, power is power, and cables and fuses go hand in hand. Electricians may often frown strangely when asked to supply a fused, 50-amp feed with cable rated at 100 amps or more, but when the reasons are explained they are usually willing to comply. The situation is analogous to using loudspeaker cables of much higher current rating than is necessary for the continuous rated power handling of the loudspeakers. The rules, in fact, are identical: to supply the loudspeaker, or the studio, with the shortest, lowest impedance cables which can be practically realised.

Impedance, for anybody unfamiliar with the term, is the mixture of resistance and reactance which is presented to an alternating current. It is the combination of the DC resistance, which can be measured by an ohm-meter, and the inductive and capacitive reactances. This can fool many people. A short, thin, closely intertwined pair of cables may possess an identical resistance to a longer, fatter, spaced-apart pair, but the former will possess a lower impedance than the latter. Inductance is an electro-magnetic phenomenon which tends to cancel as the pair of wires forming the circuit are brought closer together, because the equal and opposite magnetic fields tend to cancel. The capacitance between the cables will increase as they are brought closer together, but this is a parallel phenomenon, whereas the inductive reactance is in series with the DC resistance. In practice, at power supply and audio frequencies, and especially with cables of typical studio lengths, the capacitive effects are negligible.

From the point of view of both power wiring *and* loudspeaker cabling, cables sharing the same outer jacket, and hence which are closely physically coupled, tend towards having the lowest impedances, and hence are best suited to studio use. In many countries, earth wires are also required to lie close to the live and neutral cables. It is hard to find specific data on this, but it is believed that the impedance of the earth wires tend also to be reduced by this method of dressing, but conditions can be very variable, here. Nevertheless, obtaining the lowest possible impedance path to the earth point, from each piece of equipment, is something that should be sought. The earthing system should also follow the same rules as the live and neutral feeds, in that the earth wire should share no common length of cable with any other building earth, as shown in Figure 4.1(a)(ii).

4.2 The number of phases

Some people may have laughed at the end of the last section but one, when reading the statement that all the interconnected equipment should be on the same phase. Strangely, though, they may have laughed for different reasons. In some countries, such as the United Kingdom, anything other than such a mono-phase connection would be prohibited by regulations which state that no pieces of single phase equipment connected to different phases can be installed within the distance of the spread of a person's arms. This is so that the two pieces cannot be touched simultaneously. On the other hand, in southern Europe, I have seen 12-way plugboards, installed inside studio equipment racks, with four sockets connected to each of the three phases.[1] These were installed by licensed electricians, who wondered what I was talking about when I questioned them about it. To them, it was an absolutely normal procedure. I even recall working in a large film dubbing studio in central Paris, France, which had a multitrack tape recorder on one phase, the mixing desk on another, and the effects and power amplifiers on yet another phase. To have been able to supply one phase to the studio, large enough to supply all the equipment, would have meant re-wiring the whole mains supply system within the building, which the owner was not prepared to do.

Electricity companies supply three phases because it is a much more efficient way both to generate and to distribute electricity in large quantities. Just as multi-cylinder piston engines run smoother than their similar sized, single cylinder counterparts, so three-phase generators run smoother and more efficiently than single-phase generators, in which the electromagnetic effects of each revolution are less balanced. When power can be balanced across three phases, it requires less cable to distribute it, because any neutral cables carry composite currents which are not in-phase, and so do not produce the cable heating effect of the total current passing in a single phase, neutral 'return'. More efficient generators and distribution mean less costs to the electricity companies, and hence cheaper electricity for the users.

The United Kingdom is a relatively rich, industrial country, with plenty of electricity and a culture in which concepts of safety are deeply entrenched. It is quite normal for a single domestic dwelling (house or flat [apartment]) to have a single phase 100 amp supply. In Iberia, on the other hand, where electricity is in short supply, a whole house may be limited to a 15 amp, single-phase supply, and many commercial premises may only have a 20 amp, three-phase supply. Requests for more power can incur large installation costs and a heavy surcharge on each subsequent bill. The British requirement for a single phase supply to all interconnected pieces of audio and/or video equipment is based on the premise that as potentials of around 400 volts can exist between phases, then equipment insulation could be more likely to break down when used with such supplies, as compared to use on a single, nominally 230 volt supply. What is more, the effects of any breakdown across different phases would be potentially more lethal. The Iberian philosophy seems to be 'That is all that we have got – take it, or leave it'. The French, no doubt, explain it with the ubiquitous 'Gallic shrug'. In fact, in the USA, there are even bi-phase supplies, derived from the two half-windings of a centre-tapped transformer, giving 120 volts from each half; frequently with one half going to each of a pair of adjacent houses. Internationally, these things are far from standardised.

4.2.1 Why one phase only?

Anyhow, the science, behind all of this is *not* a variable, and it is for such reasons that a single phase supply is needed. If all the studio equipment were to be operated split across three phases, then all would be quiet as long as the three phases were supplying pure sine waves and were taking equal current. However, perfect sine waveforms on mains electricity supplies are somewhat rare, and studio equipment is not like the three balanced windings of a three-phase electric motor. It does not draw constant current, as things are being turned on and off, such as tape machine motors, or have variable current demands, such as most power amplifiers when music signals are passing through them. Current imbalances produce harmonics, and strange currents can be created in the common neutral wire. This subject is dealt with in great detail in two of the books listed in the bibliography at the end of the previous chapter. To quote from the Davis and Davis book (1997, p. 394), 'It can be seen ... that noise generated by the power system can be minimised if only a single-phase power system is used throughout the electronic equipment'. To quote from Giddings (1995, pp. 54–56), 'Another issue regarding AC power is the number of phases that are being used to drive the audio system. The ideal number is one, because of the capacitive coupling between most electronic equipment's case and the power supply within. This coupling creates a voltage fluctuation at the line frequency on the case, which is usually the ground reference for the electronics within ... Pieces of equipment on the same phase will have a similar oscillating ground-reference, so they tend to cancel. If a second piece of equipment operating on a different phase is interconnected with these, a small voltage difference exists in the ground references due to the phase differences, and can be picked up as common-mode ground noise by the input, and amplified ... The more often the signal passes between pieces of equipment powered by different phases, and the greater the gains of this equipment, the more the problem is compounded.'

Over many years now, the method of power distribution that I have used for small studios is to power all of the audio equipment from one phase, all the single-phase air conditioning, or heating, on another, and the lights, ventilation and general power (to the office etc.) on the third phase. If one phase is obviously cleaner or more stable than the others, then this one should be chosen for the audio. However, this should be periodically checked, as circumstances can change as a result of modifications in adjacent buildings.

4.3 Line filters and power conditioners

As mentioned earlier in this chapter, it can be of little use conditioning the incoming power if there are pieces of equipment *within* the system which are introducing electromagnetic interference (EMI). Nevertheless, additional filtering of the supply can do no harm. The trouble is that so many EMI or RFI (Radio Frequency Interference) filters (or suppressers) fail to clean up the problems, and merely result in dumping even more electrical hash on the earthing system. Figure 4.3 shows how this can commonly take place. In fact, the notion that uninterruptable power supplies will supply an absolutely clean supply are often very misguided. Only units with internal output voltage feedback control will ensure a clean waveform, and such units often need to be of substantially higher capacity than initially assumed, or even they may cause more problems than they solve. The problem is that they rarely have sufficiently low output impedances to ensure freedom from the production of harmonics by mechanisms such as power amplifier transient demands. Some of these things can actually *cause* computer crashes unless they are connected to individual computers without any other equipment sharing the supply. (For readers unfamiliar with these mechanisms, a careful re-reading of Figure 4.2 could be useful.)

4.4 Balanced power

Figure 4.4 shows the balanced equivalent of Figure 4.3. The drawing was

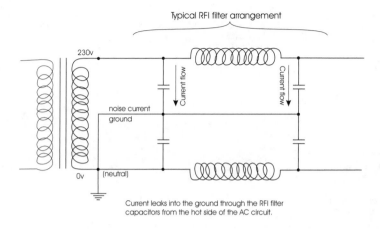

Figure 4.3 RFI filters in unbalanced circuits. Radio interference (RFI) filters can often remove much noise from the AC supply, only to dump most of it on the ground, which may be equally noise-sensitive: 'Out of the frying pan, into the fire!'

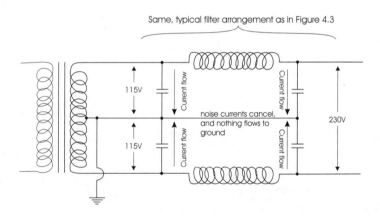

Figure 4.4 RFI filters in balanced circuits. The grounded centre tap on the AC transformer provides grounding for the AC system and balances the leakage currents to ground, thereby eliminating capacitive leakage from RFI filters as a source of objectionable grounding currents, and the consequential noise

taken from an article by Martin Glassband,[2] the author of the 1996 amendment to the United States National Electrical Code (Sections 530–70 to 530–73). It has long been known that supplying sensitive audio equipment from a balanced power source can resolve many noise and crashing problems. The essence is to use a centre-tapped transformer, supplying two supplies of opposite polarity, each of half of the total voltage. For example, a 230 V supply is replaced by a centre-tapped supply, supplying two 115 V 'halves' balanced around a centre tap which is taken to earth. Figure 4.5 shows how difficult reactive current problems can also be nullified by balancing the

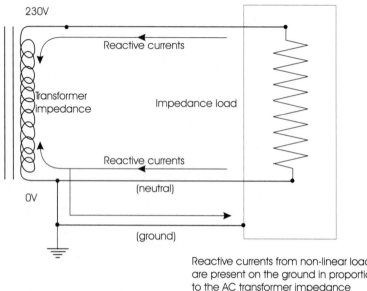

Reactive currents from non-linear loads
are present on the ground in proportion
to the AC transformer impedance

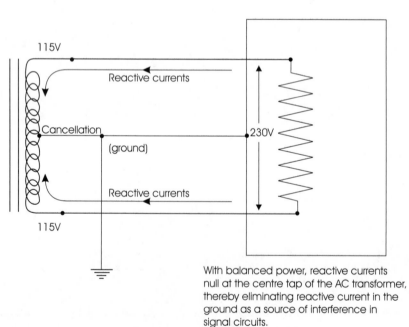

With balanced power, reactive currents
null at the centre tap of the AC transformer,
thereby eliminating reactive current in the
ground as a source of interference in
signal circuits.

Figure 4.5 Comparison of behaviour of reactive load currents in unbalanced and balanced AC systems

power feeds. In some countries, to supply the wall sockets of a building in such a way may be prohibited by local regulations, which tend to lag behind the needs of rapidly advancing technology.

In the USA the fact was recognised in the late 1980s that certain high technology systems were so sensitive that conventional electricity supplies were limiting their performance. In 1996, there was an amendment added to the US National Electrical Code which allowed audio-video and similar installations to be run from balanced power systems, as long as certain specific conditions were met. In other countries, after due explanation, it may be possible to persuade the local electrical safety people of the necessity. On the other hand, as such codes relate only to systems installed within the structures of buildings, it may be possible to use a balancing transformer as a free-standing unit, feeding the plugboards to which the audio equipment is connected. This is not something which should be undertaken without expert advice, however, because the use of isolation transformers might defeat the operation of differential circuit-breakers, which are essential safety features in many installations.

There are now many installations in which balanced power has been used, and the benefits can be quite wide-ranging. Studios with historic problems of noise have reported over 15 dB less interference pick-up. Due to the ability to null high frequency interference in a way that is almost unachievable in conventional systems, jitter has been reduced in digital circuits, allowing much cleaner audio signals. The measured reduction in jitter has sometimes been more than 30 per cent. There is also an effect on capacitive interference, because the effects of an electric field are related to the square of the voltage. A nominally 230 V balanced system can only be 115 V away from ground, compared to the 230 V of an unbalanced system. (Or 60 V instead of 120 V in some countries.) Halving the voltage to ground therefore reduces the interference strength by 2^2, or four times. Balanced supplies can be very effective; even in the most arduous of circumstances.

One reason why balanced power is so appropriate to project studios is that, unlike fully professional studios, they are very likely to have the sort of equipment mix which does not easily lend itself to optimal audio interfacing, having mixed signal levels, many balanced to unbalanced connections, and much equipment that is only marginally engineered in the first place. Power balancing is also appropriate because project studios are often sited in buildings where dedicated earths and low impedance mains supplies are hard to realise in practice. They may also be in remote areas, at the end of long power lines, or on the 21st floor of a building, where a good earth can be hard to come by. Remember, though, never to attempt *any* electrical modification without the aid of a qualified electrician. Lives are at risk!

4.5 Summary

Whilst it is way outside the scope of this book to go into details of how to make special earthing or power supply arrangements, hopefully it can at least stimulate awareness of the range of problems and solutions. Indeed, it would be totally irresponsible of me to encourage non-specialised personnel to attempt to make any changes on such safety related issues. Anybody who is particularly interested in the subject should refer to the books mentioned in the bibliography at the end of this chapter, after which they will realise just how deep and complex the subjects are.

The main point to be made in this chapter is that a good, clean mains supply and earth are not to be expected from a simple, domestic supply outlet. Many more computer crashes are caused by power problems than are ever realised, and software is often blamed when it is not at fault (although very frequently it is). Many noises are blamed on poor earths when they may actually be caused by poor mains, and vice versa. Many power conditioners do not provide the degree of cleanliness that their users believe that they will deliver, and many uninterruptable supplies are not as free from mains-borne interference as they would lead their users to expect.

Once again, domestic and semi-professional equipment may be much more prone to disturbances by poor mains supplies and earthing arrangements than their fully professional counterparts, yet they may be the very pieces of equipment which are most likely to be used in less than optimum conditions. I really wish that I could have written a chapter saying 'If this happens, do that; if that happens, do this,' but I am afraid that the subject is not so simple. What I do hope, though, is that the chapter will have given an insight into the roots of the problems, their various effects, and some suggestions that there may be lights at the ends of the tunnels. One thing which must be borne in mind, though, is that the fixing of poor mains supplies and earthing, to get the best out of semi-professional equipment, may be more expensive than buying professional equipment which may be more tolerant of poorer supply conditions.

References

1 Newell, P.R., 'Namouche', *Studio Sound*, Vol. 36, No. 12, p. 48 (December 1994)
2 Glassband, Martin, 'The Origin of Balanced Power', *Sound and Video Contractor*, pp. 54–60 (September 1997)

Bibliography

Davis, D. and Davis, C., *Sound System Engineering*, Focal Press, Boston, USA (1997)

Giddings, P., *Audio System Design and Installation*, Focal Press, Boston, USA and Oxford, UK (1995)

Giddings, P., *Audio System Design and Installation,* Howard W. Sams, Indianapolis, USA (1990)

Journal of the Audio Engineering Society', Vol. 43, No. 6, issue dedicated to papers on audio system grounding and interconnecting (June 1995)

Sound and Video Contractor, (P.O. Box 12901, Overland Park, KS 66282–2901, USA) (September 1997). Issue dedicated to power quality.

Chapter 5

Monitoring requirements

Obviously, the benefits to be gained from the application of the techniques outlined in the previous two chapters may not be fully appreciated, nor even be deemed to be necessary, unless there exist monitoring conditions of sufficient resolution to be able to hear either the problems themselves or the effects of their removal. Before going any further, therefore, perhaps we had better take a look at the requirements for adequate monitoring, which, especially in small studios, provides the one and only window through which the proceedings can be judged properly.

Despite what many people think, headphones are not a viable alternative to a good monitor loudspeaker system and a good room. Human physiology, as well as various aspects of psycho-acoustics, give rise to very great differences in perception when listening via loudspeakers or via headphones. Generally speaking, a mix done on loudspeakers, under reasonable conditions, is much more likely to representatively transfer to headphones than a headphone mix is likely to be similarly perceived via loudspeakers. The options available for providing good loudspeaker monitoring are not exactly inexpensive, but they are absolutely crucial to the achievement of good quality recordings and mixes. It is foolish to skip on the monitoring, because it is the one aspect of the studio through which all else will be judged, and the potential of much expensive equipment may never be realised unless it can be heard.

5.1 To a standard, or to a market?

The outlook on monitoring can be seen from two very different viewpoints. Do we monitor to a standard, or do we monitor to a market? With the latter, we risk a slippery slope downwards, with a difficult route to claw our way back, yet the philosophy does have its disciples. Monitoring to a standard, we need full range monitoring systems, with what is generally accepted to be the most neutral and accurate sound quality. Obtaining the desired 'sound' or balance at the lowest end of the frequency spectrum has *always* been difficult. This has often been the area which has sorted the wheat from the chaff in engineering terms. Obviously if one cannot hear the bottom two octaves, one does not have the problem, but is this *really* a professional attitude to the problem? The contrary argument to this is that the marketplace largely exists of people having music systems which do not respond to the full frequency range, so why do we need to worry unduly about it? From a commercial point of view therefore, if the mix is optimised for the equipment of the majority of people, then the greater number of the people should be pleased by the results. This *should* represent the most cost-effective approach, as not too much time would be wasted on correcting largely unnoticeable problems. The root of some of the use of less critical monitoring has stemmed from the use of computer generated sound sources, where it is often presumed that samples or internally programmed sounds are relatively free from unexpected problems at the frequency extremes, and that they will have been properly 'designed' in the first place. The use of compression and equalisation, though, can soon change this situation. Recordings which have not been referenced to high quality monitoring, however, cannot realistically pass themselves off as hi-fi.

By the less enlightened, the term 'flattery' has often been misapplied to the use of full range monitors. In monitoring use, loudspeakers which 'flattered' in the traditional sense have never been well received. This traditional sense implied that such a loudspeaker enhanced the sound, beyond the natural quality of the real sound. On a 'flattering' loudspeaker, results could be obtained which, despite sounding vibrant and exciting at the time of mixing, were disappointing when played elsewhere. With acoustic or electric instruments, it was always possible to use an actual instrument as a reference, to judge whether the monitors were neutral, disappointing, or flattering with respect to the original sound being recorded. However, I am now being confronted with engineers who refer to as flattering what I, and many of my colleagues, consider to be neutral monitors. In this instance, however, 'flattering' has been adopted in a very different context. 'Flattering' is here being used to mean 'sounding better than on my small loudspeakers at home'. In other words, full range monitoring would automatically be deemed to be 'flattering' in the 'mix to the market' approach.

Subsequent to pursuing the 'mixing to the market' philosophy on small loudspeakers only, the use of computerised sound sources, and the absence of any 'live' sound source to use as a reference has often led to the domestic situation becoming the *de facto* new reference. Increasingly, engineers are recording and mixing on arbitrary, close-field monitor systems, then judging the results on similar loudspeakers at home. Surely, this is approaching a self-fulfilling prophecy. Using what I consider to be far more accurate monitor systems, the apparent frequency range, transparency, cohesion, imaging and general quality would increase significantly, at least when listening to things which are well recorded. Some less experienced recordists tell me that this is flattering the real sound, because it is not what the music sounds like on the lower-fi, close-field systems, *or* the home systems. In other words, they are saying that they do not *want* to hear what it *really* sounds like, because that is not how they will be listening at home, and furthermore, hearing what it *really* sounds like, makes mixing more difficult. This is not a professional attitude.

If this path is pursued, then I do not hold out much hope for maintaining the more worthwhile standards of the recording industry, and the days of the audiophiles will be over. It should be remembered that we owe a great debt to the audiophiles for constantly pushing for better recordings. They keep us on our toes. Without them, who knows how far standards would have fallen. From my point of view, we should all take up the challenge that they represent, and on their 'super-fi' set-ups, they *deserve* to hear something better from a recording than would an 'average' listener. Above all, it is an outrage if their better equipment only allows them to hear problems that should have been sorted out in the studio. A clear sign of good studies and skilled engineers is when the recordings which they produce exactly track the quality of the reproduction systems: better system, better sound. *Not*, better system, more problems noticed.

5.2 Minimum standards

Personally, I do not advocate the use of a single set of small monitor loudspeakers, such as the Yamaha NS10s, as the sole studio reference. They are simply not sufficiently revealing of the details. Multitrack mixing and the use of effects processors can produce some strange and totally unwanted artifacts which may not be detected on 'average' quality loudspeakers. I realise that good monitors are relatively expensive, but they are, after all, the only means of assessing tonal balance, non-linear distortions, phase anomalies, transient accuracy, spaciousness, and all the other characteristics which must be monitored if high quality music recordings are to result.

In 1999 prices, a reasonable set of monitors would cost around $1500–$2000. Such units would be similar to the Genelic 1030As and the Quested F11s, though the low frequencies responses are still not sufficient for true, full-range monitoring. Both of these are active units, that is, they have their own built-in, low-level crossovers, and separate power amplifiers for each drive unit. For around the same price, ATC SCM10s could be bought, but they would tend to require the same amount spending again on a good quality power amplifier. There simply is no way to make small loudspeakers which go loud and have an extended frequency range without the use of low sensitivity drive units and high power amplifiers. The laws of electroacoustics simply preclude it. The reasons why are way outside the scope of this book, but for anybody interested in the subject, the books mentioned in the bibliography contain the details.

Remember, it is unreasonable to use the term 'control room monitoring' without due thought being given to the first and last words. They imply quality control and a reference. Neither of these functions can be achieved by loudspeakers or amplifiers which are not at least as good as the better types of domestic reproduction equipment, and studio monitors must also take much more punishment from the auditioning of single instruments, and accidental overloads. To make such systems is not an easy task, and this is also a reason for the relatively high prices of studio monitors of acceptable quality.

5.3 A better way

So, after having discussed so many of the limitations of oft used monitor systems, precisely what sort of system should we be using? In the introduction to my first book[1] I quoted George Massenburg from an interview in *EQ* magazine in June 1993. He said, 'I believe that there are no ultimate reference monitors, and no "golden ears" to tell you that there are. The standards may depend on circumstances. For one individual a monitor either works or it doesn't ... Much may be lost when one relies on an outsider's judgements and recommendations.' For anybody unfamiliar with the name, George is one of the world's most respected recording engineers and producers, and was the inventor of the parametric equaliser. He is the founder of GML, the producer of ultra-high-end studio equipment, and is closely involved with Tonmeister work at McGill University, Montreal, Canada.

We all have different auditory systems, and we all have different hierarchies of exactly what is important to us in any piece of music. We do not all gain our enjoyment from the same aspects of any given piece of music, even if we may appear to enjoy it to an equal overall degree. Furthermore, different types of music and different techniques of recording all tend to demand different priorities in terms of loudspeaker performance. Perhaps we should take a short diversion from track, here, and take a look at what we can glean from what we can measure, and how that relates to what we perceive.

5.4 Loudspeaker assessment

The problem is how to do authoritative and repeatable tests in the objective sense which clearly relate to widespread subjective perceptions. What, for example, can we measure about an NS10 which relates to its decade-and-a-half of popularity in recording circles? Howsoever these tests would be done, they would be unlikely to lead to any outright winners, as, even in a purely objective world, different loudspeaker responses tend to suit different circumstances. Subjectively, user expectations and preferences will also vary, as will the demands of different types of music or recording techniques. In fact, some of these requirements appear to be mutually exclusive, and will probably remain so as long as we do not have perfect sound reproduction systems. What we do have, however, are *some* performance 'truths' which are relatively incontrovertible in their objective/subjective relationships. If we can show how the perceived characteristics of some groups of loudspeakers relate to their objective groupings, then hopefully we can show the connections, and thus provide a grounding which will help to illuminate the wider picture in a much clearer light. Before moving on to the discussion of our 'truths', though, perhaps we should look at a list of *realities,* which the marketing departments of a very large number of loudspeaker manufacturers would rather that we should forget. In this 'market forces'-led world, it does well to remind ourselves, from time to time, about some of the more truly professional points of view.

5.4.1 Basic realities

- There is no loudspeaker which is optimal for all rooms.
- There is no loudspeaker which is optimal for all music.
- There is no loudspeaker which can perform optimally in an acoustically bad room.
- High definition full-range monitor systems with low distortion, good time-response accuracy, and very uniform amplitude and phase responses, cannot, as yet, be made cheaply. $5000 per pair would seem to be about the minimum.

- Computer-aided design (CAD) does not mean that products derived from such techniques are necessarily superior to non-computer-aided designs.
- New technology cone materials are not always subjectively superior to older ones.
- Too many loudspeakers are being marketed to what is now such a computer-orientated industry that the use of CAD technology is itself a selling point. I have seen people buy loudspeakers on the strength of their specifications alone, without auditioning them, which is ludicrous.
- The Quad Electrostatic Loudspeaker of 1957 can still give an account of itself which the last 40 years of development have not been able to significantly better, at least not within its axial SPL capability. (And, for similar reasons, what microphone can put to shame a Neuman M49 from almost 50 years ago?)

Studio designers must bear many of these facts in mind to achieve the best results from any given set of circumstances. Now let us take some more individual aspects of performance into consideration.

5.4.2 Subjective/objective truths

Amplitude responses
It is generally considered that the single most important aspect of the perceived 'accuracy' of a loudspeaker, or an entire monitor system, is that it should provide a uniform acoustic pressure amplitude response (or frequency response in more common jargon) at the listening position.

Phase responses
Largely due to work by Ohm and Helmholtz in the last century, it has been a widespread belief that phase response uniformity was not particularly critical. Unfortunately, the above work was carried out using sine waves, which bear little relation to music signals. This subject I have dealt with, at some length, in another book,[1] but briefly, the phase response, in combination with the pressure amplitude response, defines the time response, and the three are inextricably linked by the Fourier Transform. In the days of analogue dominated recording, phase inaccuracies inherent in the recording process rendered high degrees of phase accuracy in the monitoring systems to be somewhat less important, but digital recording has put a new emphasis on the phase accuracy. In order to perceive very low level detail in digital recording, modern control rooms tend to be acoustically rather dead, and these conditions happen to be exactly where phase inaccuracy becomes more noticeable.[2]

Time responses
In 1989, Keith Holland and I wrote a paper which I presented to the Reproduced Sound Conference of the UK Institute of Acoustics,[3] in which we asked for more manufacturers to publish the impulse responses of their systems. The integration of this, the step-function response, is perhaps a more visually representative form, and step functions themselves are actually very realistic test signals (see Figure 5.1). Some delegates questioned whether this was not 're-inventing the wheel' as waveform analysis was an old technique

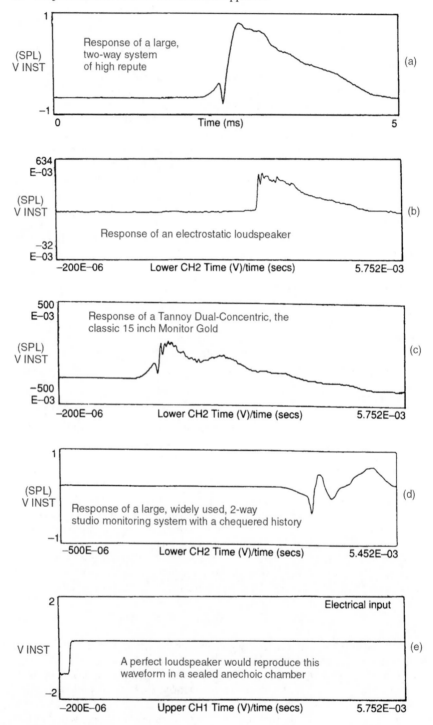

Figure 5.1 Full-range step function responses

that had been rejected as showing little relevance to perceived performance. The 1989 argument was that, in the past, the time responses of all but a very few loudspeakers (such as the Quad Electrostatic, which was instrumental in the conception of the paper) were simply too far away from time accuracy to be compared on such a basis. In 1999 people *are* publishing time responses as more attention is generally now paid to getting them better, largely in response to the demands of digital recording. A very great deal of the natural sound character of an instrument is dependent upon the integrity of the leading edge of its waveform, and this can only be preserved if a loudspeaker system maintains the time integrity of the signal. Transparency and openness are also very dependent on the accuracy of a loudspeaker's time response.

Non-linear distortions
Harmonic and intermodulation distortions produce sounds which are not a part of the input signal. Higher harmonics generally lead to harshness and aggression in the sound, making any given sound pressure level (SPL) seem subjectively louder, but masking some of the low level detail in the signal. It is also difficult to produce an open, expansive, transparent sound with significant levels of non-linear distortion. Second harmonic distortion can produce richness and warmth in the sound, but if this exists in the monitor loudspeakers, it is not necessarily on the recording, so will not travel with it. If such a sound is wanted, then one should use some valve (tube) microphones or valve electronic processors, and *record* the sound of the harmonics. It is no good just to listen to it on the recording/mixing monitors. On the other hand, people who like these sounds at home are at liberty to buy loudspeakers which produce 'pleasant' distortion if they so choose, but that is for their own entertainment, not for quality control listening.

Undefinables
Over a period of years of listening, it has become evident that some 'advanced' types of loudspeakers are lacking in low-level detail and general openness, whilst seeming to be almost exemplary in other respects. After reading an AES paper by David Clark[4] in which he discussed stiction effects in low frequency (LF) motor systems (drive units), I wondered if the problem could be in the suspension systems, whose inherent losses require a finite minimum drive signal to get them moving. After speaking to David, he convinced me that these were purely LF problems, and that the frequencies of interest for transparency and openness were above the affected range. I was partly led astray by associating these problems with some new suspension systems, but Martin Colloms points out in his book *High Performance Loudspeakers*[5] that he believes the problem to lie in some of the new cone materials, some of which, co-incidentally, were introduced with some of the new types of suspension systems. He believes that some of the new, very rigid, high internal loss cones exhibit hysteresis losses which are very hard to quantify, yet which can drastically affect the perceived naturalness of their response. Some of this problem may have arisen as a result of the over-use of computer-aided design, where solving one engineering problem (such as creating ultra-damped cones) has, itself, caused other problems, but the computers have never heard them. In listening tests, designers can, all too frequently, be drawn into listening so specifically in the area of interest that they often

fail to notice negative side-effects which result from any 'improvement'. Technical 'advances' must be very carefully subjectively assessed. They do not always bring the *overall* advance in audible performance that they suggest from their *specific* performance improvements.

5.5 Which way to go?

In some of these areas, we are still in troubled waters when using direct measurements, and perhaps the best insight into the origins of such problems can be gained from the grouping of similar materials or similar design concepts to similar sonic performances. Careful listening tests in acoustically controlled environments are maybe the only reliable way to go about the pursuit of the facts. Although this is time-consuming and costly, it is, nonetheless, perhaps the only way that we can currently find the sources of some anomalies in the objective/subjective comparisons. However, in order to do the comparisons, we must have the objective measurement results.

There would seem to be little doubt that one of the greatest problems facing the designers of loudspeakers for critical listening purposes is the human perception factor. Over the years, loudspeaker design has largely advanced in bursts, but these bursts of development have frequently been due to new audiological/psychoacoustic information becoming available to the designers, as opposed to the use of 'revolutionary' new materials. In fact, the latter have frequently not lived up to their technological promises once a period of critical listening has been completed. We are still a long way from truly being able to specify or quantify many of the parameters which impart on one loudspeaker that 'special something' which gives it an edge in terms of acceptance.

Furthermore, the situation is complicated by many people, even experienced professionals, having different ideas of what they expect from a loudspeaker; especially in terms of their hierarchy of performance demands. This situation exists with other subjective senses, though, and is not exclusively an audio-related phenomenon. Kodak, Fuji, Konica and other film manufacturers all have strongly partisan followings. Personally, I am a Kodak user, because Kodak films and processing give *me* photographs which *I* see as being the best representation of what I saw at the time of taking the pictures. No Fuji fanatic is going to convince me otherwise, yet equally, I realise that if Fuji fans see the beauty that *they* are looking for in the blues and greens, then far be it from me to argue with them.

This situation exists despite the fact that a photograph can be held up in front of the original object, and, given the same light as at the time that the photograph was taken, the colours can be directly compared. It also exists despite the fact that the optic nerves carry a relatively complete signal to the brain, and these signals can be measured, from person to person, and the differences can be assessed. On the other hand, no sound reproduction system can ever be compared, side by side, with an actual event, except perhaps in an anechoic chamber, but even this is fraught with problems. *No* loudspeaker ever radiates sound in the manner of an acoustic instrument, not even for something simple, like a solo flute, or even a triangle. (No instrument emits all its high frequencies out of one point, as do most loudspeakers.) The sound field distortion which is thus produced means that each one of us will hear different differences between the live and recorded sounds. The reason for this is that human pinnae (outer ears) are as individual as fingerprints. The way that they collect the sound is therefore unique to each one of us, and sound field differences will, in turn, be differently perceived from one person to another.

5.6 No ultimate resolution?

To make matters worse, unlike the optic nerves, which carry a comprehensive electrical analogue of what the eye is looking at, there is no such parallel in the auditory system. Nerves from the inner ear disappear into about half a dozen different parts of the brain, and there is nowhere where doctors can measure a complete electrical signal relating to what we are hearing. Perhaps the final blow comes from the fact that our perception of music is highly unstable in terms of how it can be drastically changed by our moods. These are the sort of problems which called for the creating of a whole branch of science called psycho-acoustics. It exists because of the extraordinarily complex ear/brain relationships, and often attempts to define areas which allow equipment designers to take most advantage from what the brain needs to receive for a maximisation of realism with a minimum of complications. This is the only path which we can currently follow if we are seeking the highest fidelity of reproduction that we can achieve, because the cost of truly recreating an original sound field would be enormous.

The late Michael Gerzon, about 10 years ago, suggested that if we wanted to reproduce an accurate sound field of a musical instrument, then we would have to surround it by about a million microphones, going to a million-track recording device, and reproduce the recording via a million amplifiers and loudspeakers. Even *that* system would assume perfect microphones, and that the one million perfect microphones did not reflect sound back towards the instrument or the other microphones. In other words, we cannot do it.

The above paragraphs should have made the point abundantly clear that loudspeaker reproduction of sound is not even close to recreating a 'real' sound, except, perhaps, if that original sound was itself generated by a loudspeaker in anechoic conditions, which is a rather unlikely set of circumstances for a realistic music recording process. Given the limitations, it should also be realised that objective assessment of loudspeakers can only go a certain way towards describing the subjectively perceived aspects of their performance. The fact was highlighted over 20 years ago by another long lamented member of the doyens of audio, Richard Heyser, who said that in order to fully enjoy the intended *illusion* of a recording, it is necessary to willingly suspend one's belief in reality. He continued, '*All* recording and reproduction via *two* loudspeakers is illusory: it will take powerful signal processing systems and multi-loudspeaker technology before we will ever be likely to see the three main groupings of classical, rock, and cinema in absolute agreement.' With such visionaries as Michael Gerzon and Richard Heyser seeming to agree completely on such an issue, it would either take another genius, of even greater understanding, or an utter fool, to argue with them.

By performing many objective measurements, however, it is possible to look for groupings which can then be correlated with the subjective opinions

of the loudspeaker users. By this means it may be possible to isolate performance characteristics which lead to certain sonic strengths and weaknesses,
and, given time, more will surely be learned about the relationships. For the
time being, though, we can only do the best that we can with the knowledge
and technology that we have available to us. For now, the most important
thing is to be aware of the current limitations which we face and deal with
them appropriately.

So, what we seem to need for studio monitoring purposes are low distortion, low coloration, fast transient response loudspeakers, with minimum
phase distortion, a smooth and preferably extended frequency response, and
adequate output capability for the purposes for which they will be used.

Note, here, that I make no mention of power handling, because, on its own,
it is a meaningless specification. What we are really interested in is the
maximum acoustic output, or, 'how loud they will go'. To highlight this, take
the case of two studio monitors such as the ATC SCM10, and the UREI 815.
Both are purpose-designed studio monitor loudspeakers, but they are enormously different in sensitivity. One watt into the SCM10 will produce an
output of 81 dB at one metre distance, but 1 watt into the 815 will produce
103 dB at one metre. For every 3 dB increase in signal level, an amplifier will
be asked to deliver twice as much power, so for the 22 dB of sensitivity difference between the two loudspeakers under discussion, here, the difference
in required amplifier power to produce the same output from each loudspeaker can be seen from the following table:

Table 5.1

dB increase	Multiplication of power
3 dB	× 2
6 dB	× 4
9 dB	× 8
12 dB	× 16
15 dB	× 32
18 dB	× 64
21 dB	× 128
24 dB	× 256

Twenty-two decibels (dB) would therefore represent a power increase of
somewhere over 150 times. It therefore follows that one watt into the UREI
815 would produce equal loudness at an equal distance to somewhere
between 150 and 200 watts into the ATC SCM10s.

The 815 is a large, double 15 inch woofer design, whereas the SCM10 is a
small, close-field monitor. As mentioned in Section 5.2, the physics of loudspeaker design dictate that small loudspeakers cannot be made as efficient as
large ones, well, not for equal bandwidth, that is. However, once the above
facts are appreciated, it will become clear why 'power handling capacity'
means nothing if it is not related to sensitivity. What it does mean, though, is
that small, high quality, wide bandwidth loudspeakers may well need enor-

mous amplifiers to drive them, and unless those amplifiers are of sufficiently high quality, and of adequate transient output current rating, the potential of the loudspeaker performance may never be realised. The upshot is that when paying $1,500 for a good set of non-powered loudspeakers, one should be prepared to spend the same amount, again, for a suitable amplifier to drive them. This helps to explain why small, high quality, wide band width, self-powered monitor loudspeakers may cost $3000 or so. Good, small, passive loudspeakers will typically need amplifiers of around 300 watts per channel into the appropriate impedance if a good transient response and transparent sound are required; and for monitoring purposes, they are.

5.7 Self-powered, or passive?

Many small monitor loudspeakers are now available with built-in amplifiers. These have the advantage that they use amplifiers which are considered by the manufacturers to be appropriate for the frequency ranges of the individual drivers, and they reduce the length of loudspeaker cables to an absolute minimum. The use of low-level, electronic crossovers, before the amplifiers, has many benefits over the use of passive, high level crossover, after the amplifiers. These include much more precise control over the loudspeaker cone movements, more precise matching of the electronic filter responses to the actual driver responses, less of a tendency for overloads to damage tweeters, no potential quality loss with loudspeaker cables, lower levels of distortion (particularly at high levels) and more flexibility in the precise matching of drive unit sensitivities. Bearing in mind what was said in the last section about the cost of suitable amplifiers for good, low sensitivity, passively crossed-over loudspeakers, the cost factor does not really count against active monitors. On the drawback side therefore, the only real point worth considering is that one is stuck with the amplifiers supplied. Marketing, these days, has a stranglehold on manufacturer and designers, so it is probable that in most cases, the amplifiers have been designed to a best compromise on the cost/quality curves, and it is probable, also, that if competitive costs were not a factor *vis-à-vis* other manufacturers of similar products, better amplifiers could be used. However, one has to be reasonable about this, as users also have the option of using $12,000 Krell amplifiers on their NS10s, but few do, as almost everybody needs to draw a line somewhere.

5.8 Amplifiers

The differences in sound between different amplifiers can be quite significant. Passively crossed-over loudspeakers can be quite difficult loads to drive, and amplifiers, in general, have traditionally tended to have been tested into simple, resistive loads. Steady state output powers and very high figures for damping factors tend to look good on paper, as do lots of zeros after the decimal point in distortion figures. In reality, though, ultra low distortion and very high damping factors can sometimes be achieved by means which may not necessarily be conducive to the best overall sonic performance.[6] For monitoring purposes, low distortion is essential, but whether an amplifier with

0.001 per cent total harmonic distortion is sonically better than one with 0.07 per cent, which is seventy times greater, may depend on many other performance factors. One of the aspects of amplifier performance which is of great concern in monitoring situations is the ability to supply very large instantaneous currents. Despite all the advances of modern technology, this requirement still seems to require a large, heavy, conventional power supply, or a switch-mode supply with a higher capacity than would seem to be obviously necessary, because you cannot draw more current at the loudspeaker output terminals than the power supply can deliver. Due to reactance in the loudspeakers, which means that the voltages and currents may not be in phase, current requirements can be much greater than the simple $W = I^2R$ calculations would suggest, as previously discussed in Figure 4.2.

5.9 Achieving the objective

There are three basic functions which studio monitors need to perform. They must function as recording monitors, which need to inspire the musicians at the time of their musical creation; they must perform a quality control function, so that any problems are heard first in the studio; and they need to be usable to find a musical balance during mixing. This can be a tall order for any one pair of loudspeakers, though in the domain of project studios, it is not unusual to find one set of monitors which have been found to be a workable compromise. However, as compromises, by definition, they cannot be optimum in all circumstances.

The flattest responses over the widest frequency range can only be achieved in an anechoic chamber (which is why such chambers are used for measurements), or in the open air. The next best results can be achieved by mounting the loudspeakers flush in the front wall of a well damped room, which is the case for the main monitors in most professional studios. Sound waves expand like the ripples produced when a stone is dropped into a pond, and Figure 5.2 shows how this propagation develops, in anechoic chambers and in more reflective spaces, both for flush mounted and free-standing loudspeakers. This represents 180° and 360° radiation, respectively. In anechoic chambers, it can be seen that the propagation towards the listening position is unaffected in either case, but in reflective rooms, the free-standing loudspeakers will produce more reflexions than flush-mounted loudspeakers. This is due to the 360° radiation producing twice the number of reflexions as 180° radiation, thus disturbing the response at the listening position to a greater degree.

The popularity of 'near-field', or more correctly close-field, monitoring (near-field has some quite specific, acoustic meanings which do not necessarily apply to the way in which small loudspeakers are used in control rooms) has been because the higher the ratio of direct to reflected sound, the less the room-induced disturbance will be. Of course, the quantity of reflected energy will be a function of the room acoustics, the control of which is the domain of the studio designers and acoustics engineers. In brief, though, early reflected energy should be kept to a minimum; that is, energy returning to the ears within 30 or 40 milliseconds of the direct signal. What happens after that is a matter of which control room philosophy one wishes to follow. Nevertheless, whichever philosophy is followed, the requirements will still be

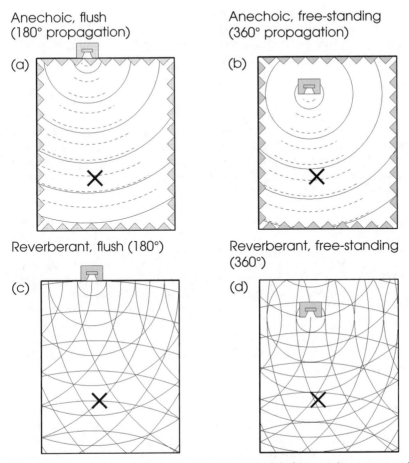

Anechoic, flush
(180° propagation)

(a)

Anechoic, free-standing
(360° propagation)

(b)

Reverberant, flush (180°)

(c)

Reverberant, free-standing
(360°)

(d)

At point X in the anechoic room, the low and high frequencies are perceived equally for both flush mounted and free-standing loudspeakers. However, in a more reverberant room the low frequencies rapidly become confused. For the free-standing loudspeaker, the confusion sets in more rapidly than for the flush mounted loudspeaker, as can be seen from the greater density of wavefront lines in (d). As the high frequencies radiate more directionally, they are not significantly affected by the loudspeaker mounting, though they will be more generally confused at point X in the reverberant room than in the anechoic room.
————— omni-directional low frequencies
-------------- directional high frequencies

Figure 5.2 Comparison of loudspeaker sitings in anechoic and reverberant spaces

to produce the smoothest, cleanest response, so that the recording staff can properly assess what is on the tape.

5.9.1 Balanced responses

At the high frequency extremes, particularly in the 'last' octave or so from say 8 kHz to 20 kHz, some different philosophies exist side by side. A large proportion of purpose designed studio monitor systems have intentional roll-offs, typically being 4–10 dB down at 20 kHz, following a gentle slope from 6 or 8 kHz. Figure 5.3 shows the responses of three monitor loudspeakers which have clearly been designed in accordance with this philosophy. Loudspeakers designed for domestic hi-fi more usually have responses continuing more uniformly to 20 kHz. There are several reasons why many studio monitor systems incorporate this high frequency roll-off. Human beings tend to perceive more subjective low and high frequencies, proportionate to the mid-frequencies, when the overall level of loudness is increased. This fact is clearly illustrated in Figure 5.4. For many reasons, studio monitoring tends to take place at somewhat higher SPLs than typical domestic listening, and when large loudspeakers with flat high frequency (HF) responses are used at moderately high levels, they tend to become unnaturally 'toppy' in their tonal balance. Furthermore, high levels of high frequencies can become very tiring when a person is subjected to them for long periods of time. There are, however, two important reasons for domestic loudspeakers having rather more HF responses as compared to studio monitors. Firstly, in many domestic rooms in which hi-fi systems will be placed there is more high frequency absorption (relative to mid and low frequencies) than in most studios, due to more soft furnishings and less hard-surfaced equipment. Secondly, domestic listening tends to be done from positions greater than the critical distance, (which is where the strengths of the direct and reflected fields are equal) and it is in the reflected field where the high frequency absorption in the room is most noticed. Domestic rooms therefore tend to produce less high frequencies in the overall, perceived response than is the case with purpose-designed studios.

Many years of experience has led to a great number of engineers opting for a gentle roll-off of the high frequency response of large studio monitoring loudspeaker systems, believing that such a curve yields the most natural results for the tapes leaving the studio. The room acoustic questions will be discussed, further, in Chapters 10 and 11.

5.10 An inverse relationship

In far too many instances, the monitor systems used in private project or semi-professional commercial studios are seen as not being in the direct signal chain, so, in many cases, as little as possible is spent on them. Yet without monitors of a highly revealing nature, there is absolutely no hope of achieving reliably consistent and desired results, and this is especially true when using equipment which is marginal in its performance. With semi-professional as opposed to fully professional equipment, aiming for the general target area is not good enough, because the noise/headroom/distortion windows are often very narrow. Only the bull's eye, *every time,* will allow high quality results to be consistently achieved. Great care must be taken at each stage of the recording process, and good monitoring is the only way to ensure that quality is being maintained. Unfortunately, and somewhat ironically, when one is less certain about the sonic integrity of the recording chain, the more critical one must be in the assessment of it. High quality monitoring has *not* followed the same reductions in the cost/performance ratios as the

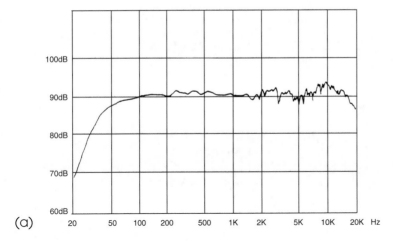

(a)

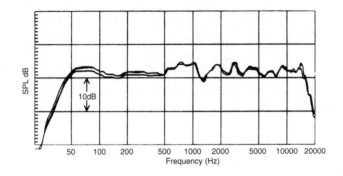

(b)

Figure 5.3 (a–d) High frequency roll-offs and general presentation of results. Responses of three small monitoring loudspeakers, showing typical high-frequency roll-off. Plot (a): the level at 20 kHz is about 7 dB down on the 10 kHz level. It should be noted, though, that the greater smoothness of the plot at the lower frequencies is probably not an indication of a very smooth LF response. The smoothing is more likely to be a result of proportionally less measurement resolution as the frequency drops. Many current measurement systems employ the use of discrete frequencies, typically at 20 Hz spacings. In the octave 1 kHz to 2 kHz there would thus be 51 measurement points, but in the octave 20 Hz to 40 Hz, there would be only two! Published responses are often very deceptive.

In plot (b), a high frequency roll-off is also apparent, but here the 20 kHz level is about 15 dB down on the level at 10 kHz. This is despite the fact that the 10 kHz level is maintained to a higher frequency than in plot (a). In this instance, the manufacturer has overlaid three plots, at 10 dB steps in output. This shows the power compression at higher levels, where voice coil heating increases its resistance, and less current can thus be drawn for any given voltage. As moving coil loudspeakers are essentially current-driven devices, a reduction in current produces a corresponding reduction in output.

Plots (a) and (b) are interesting to compare in other ways. Plot (a) shows a much more gradual HF roll-off than plot (b). In general, abrupt roll-offs do not bode well for sonic neutrality. However, plot (a) shows a considerable lift at 10 kHz, which *could* indicate an unnatural brightness. The sub 1 kHz portion of the plot is not really believable, as it is much too smooth but there is nothing to prevent manufacturers from using inappropriate measuring techniques, as there is no international convention on published responses. One of the above loudspeakers was of European origin, and the other originated in the USA.

(c)

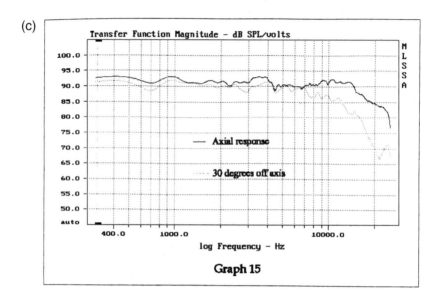

Graph 15

(d)

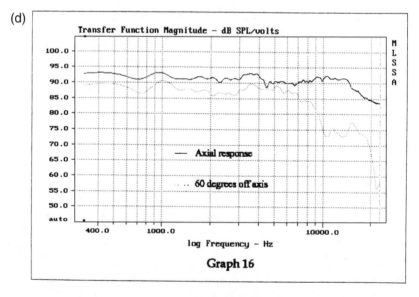

Graph 16

Figure 5.3 *cont.* Plots (c) and (d) show the axial response, plus the 30 degree and 60 degree off-axis responses of a very high quality loudspeaker in a well-controlled room. Again, the 20 kHz response is about 7 dB down on the 10 kHz response. The overall axial response is exceptionally smooth, as are the off-axis responses, which is quite rare. The high frequency roll-off is also very gradual, and is not preceded by any peaking, as is the case with plot (a). Note that this is a MLSSA measurement, which uses discrete measurement points. In this case, however, the graph has been curtailed at 400 Hz, as there is little point in showing invalid data, such as the sub-400 Hz region of plot 'a'. These plots are totally believable. The loudspeaker in question was an ATC SCM 100

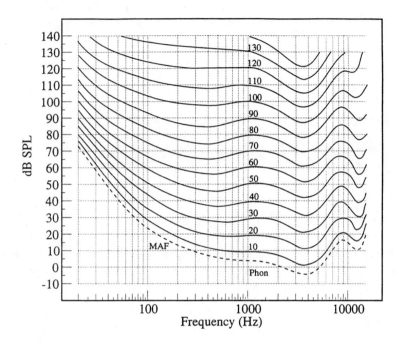

Figure 5.4 Equal loudness contours. These contours are the Robinson–Dadson curves, which updated the earlier work by Fletcher and Munson. It can be clearly seen that, at different sound pressure levels, the proportions of the low, mid and high frequencies required to give a sensation of equal loudness vary considerably. From the 100 phon curve it can be seen that 100 kHz and 10 kHz each require about 4 dB more SPL than at 1 kHz if all three are to be perceived as being equally loud. However, on the 40 phon curve 10 kHz would need to be 7 dB higher than 1 kHz, and 100 Hz about 11 dB higher, if the three frequencies were to be perceived as being equally loud. The MAF curve represents the minimum audible field, which is a statistically derived 'just above zero'. The actual zero is, by definition, imperceptible, and varies with source direct monaural or binaural listening, etc.

technological developments have brought to the main recording chain. One unfortunate fact about recording equipment is that the prices of suitable loud-speakers and microphones will not scale, in budget terms, with much of the electronic equipment on a performance to cost basis. Suitable monitors for semi-professional use do not, therefore, tend to scale in price *vis-à-vis* recording machines or mixing consoles. The big twist, though, is that to be sure of the quality of any recordings, one tends to need better monitoring systems for semi-professional chains than would be absolutely necessary if one had a recording chain of the finest components and equipment. I recently read an article about a very famous engineer/producer who announced that he now *only* used NS10s for mixing. Perhaps he does, but it would be super-foolish to try to copy his policy unless you, also, could afford to use the world's finest microphones and other recording equipment, in the world's finest studios, with the world's greatest musicians, *and* have 45 years of recording experience behind you. All our aforementioned producer is doing is balancing musical instruments. The rest he is taking for granted because of the super-everything surrounding him, and knowing that the multitracks are well recorded. Do not be fooled by reading such articles into believing that lesser mortals can get away with such nonsenses. Especially not in project studios.

5.11 Getting it right

Remember also that monitors are *not* hi-fi loudspeakers; they are specialist devices to reveal things in the sound which need to be judged and dealt with before they ever reach a pair of hi-fi loudspeakers. If top of the range, super-hi-fi loudspeakers are used, then most of them would not be sufficiently damage tolerant to withstand a bass drum, accidentally soloed with the master solo level set 20 dB too high. If medium quality hi-fi speakers and amplifiers are used, then there is no way of knowing about the subtleties which would only be exposed by the more expensive systems. After all, it is quite reasonable for the owners of expensive hi-fi equipment to expect that reputable record companies are issuing high quality recordings. To allow the top of the line hi-fi enthusiasts to be the first to notice that many recordings are sub-standard is an insult to the whole recording industry. There is of course the argument that a finished recording is an artistic entity in itself, and it *is* what it *is*, faults and all; but if the faults are there due to poor recording, then that is an insult to the record buying public who indirectly provide our food. What is more, many of the sub-standard recordings are not sold any more cheaply than the best ones, and foisting off sub-standard product at full price is an insult to the musicians and recordists who *do* try harder to maximise recorded quality.

With good monitoring conditions, quality degradations will become obvious as they occur, and the recordists' attention will be immediately called to the fact. Almost certainly, if something is going to sound undesirable to the purchasers of the finished product, then it is also unlikely to be considered acceptable to the people making the recording, so they will automatically, and perhaps even without consciously realising it, track down the source of the degradation and make whatever adjustments are necessary. With clear monitoring, life simply becomes so much easier. In some ways there is more work involved, as one becomes aware of more details which need attending to, but the *doubts* are removed from the process. There is a satisfaction in hearing things going down as intended, and knowing that one can be confident that what is being heard is honest; it will not risk being shown up later as wanting when played on a top class system. What is more, when one hears the whole process unfolding on good monitors, any adjustments made in the recording chain are heard clearly, and much knowledge is soon gained about the strength and weaknesses of each piece of equipment. Noises are heard, distortions are heard, equalisation is easier to judge, reverbs are easier to judge, and the correct operating windows of each piece of equipment are clearly obvious. To anybody of reasonable intelligence and with an aptitude for music and recording, good monitors can be as good a teacher as any human tutor.

5.12 Effects of room acoustics

Most studios based on the type of equipment which we are referring to here are in small rooms. Only rarely are the rooms treated acoustically in any serious manner. Taking any monitor system into a reverberant bathroom or empty, panelled room, will rapidly reduce its ability to resolve fine detail; things will become messy and confused. To a lesser degree, however, any untreated or poorly treated rooms will cause further confusion. It is relatively easy to deaden down the mid to high frequencies, but low frequency control in small rooms has always been a problem. If the mid and high frequencies are *not* reasonably absorbed, then they will blur the stereo imaging, colour the sound, and mask much detail by filling in the spaces in the sound with the reverberation and reflections which hang on in the room. If they *are* absorbed, but the low frequencies continue to resonate around the room, the ear will sum the power of all of the direct and reflected sounds, which will tend to mean that there is a perception of more bass, because the bass is all that is left in the reflected/reverberant character of the room. In turn, as well as creating a more muddy, boomy bass sound, mixes done in such a room may be short on bass, because the person mixing the music believed that there was more bass in the mixes than there actually was. Figure 5.5 shows an example of how poor room acoustics can cause problems in the response at the ear.

Recently, some progress has been made in the low frequency control of small rooms. The techniques make the rooms acoustically larger than they actually are. Some of the developments have come out of research being done in conjunction with Tom Hidley, who had been building control rooms, notably the BOP TV studios in Bophutatswana, South Africa, which were designed to have very low reverberation times, even at frequencies below 10 Hz. The control rooms were huge, over 16 metres front to back, but the methods used scale reasonably well, so control down to 50 Hz began to seem

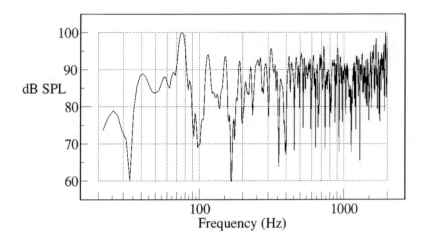

Figure 5.5 Response of a poorly treated room. This response would not be untypical of many project studio control rooms. The above plot shows the room disturbance on a loudspeaker measuring very flat indeed in an anechoic chamber.

feasible in rooms of only about 3 metres front to back. Given that many rooms of the semi-professional/private project type being discussed here have monitors which rarely respond much below 40 or 50 Hz, even with the very high quality small monitors, then suddenly, highly controlled rooms over the full frequency range of the monitors becomes a viable proposition. I have built many rooms using these techniques, which Chapter 11 explains in more detail, and the results have been quite spectacular. By using such techniques, great improvements can be made in the monitoring conditions of small rooms, and some quite remarkable 'clean-ups' can be done for a cost of around $5000 in 1999 prices.

Whilst this may still not seem cheap in domestic studio terms, what soon becomes *very* apparent is that by spending, say, $8000 on a first rate monitor system and good acoustic control, many old problems can suddenly be revealed in great detail. Once problems are easily identifiable, they can usually be more easily rectified. A great deal of the guesswork is taken out of the recording process, so work becomes quicker and more efficient. What is more, the de-bugged systems can suddenly begin to produce the sort of results which the owners previously thought could only be achieved by expensive upgrading and replacing of the equipment. Once you can hear precisely what any equipment is doing, its capabilities become obvious, so it is much easier to operate everything at its optimum levels and settings. Suddenly $8,000 does not seem that much when studios begin producing recordings which the owners only thought would be possible after spending $30,000 on new equipment. The point is worth considering carefully.

In essence, rooms can be fitted with free-hanging, highly damped panels, positioned to break up the room modes. The air gaps behind the panels are lined with fibrous absorbent felt or Rockwool, to provide good damping of the panels. The panels themselves are also treated with fibrous absorbent to kill any higher frequency reflexions, and where possible, a series of angled, deadsheet curtains are arranged in front of the panels to help to destroy the low frequency wavefronts. Figure 5.6 shows the general plans of such a construction. The ceilings are fitted with a series of sloping, full width panels, either of deadsheets or damped resonator panels, or both, and where possible a complete membrane such as of 'Acoustica Integral' 'PKB2' (which is a composite of a $3.5 \, kg/m^2$ deadsheet bonded to a $40 \, kg/m^3$, flame retardant felt) is fitted above. In all cases, the floors are hard, in order to provide some natural life to the spoken voice within the room. Figure 5.7 shows the control room whose general principles were illustrated in Figure 5.6 – Studio Dobrolyot, in St Petersburg, Russia, built in a room of 6 metres by 4 metres, equipped with Alesis ADATs, an AMEK 'Big' mixing console and Reflexion Arts monitors.

Deadsheets are usually plasticised sheets, somewhat similar to bituminous roofing felt, which act as semi-limp membranes. They are commonly available in densities from around 3 to $15 \, kg/m^2$, and are used for acoustic control. The heavier grades tend to be used when the containment of sound is required, such as in sound isolation (soundproofing) work, and the lighter grades for acoustic control applications, such as reflexion and mode control within rooms. 'Revac' is a brand commonly used in the UK, and a variety known as S.L.A.M. (Semi-Limp Acoustic Membrane) is available from Denmark. The products from Acoustica Integral SA, from Barcelona, Spain,

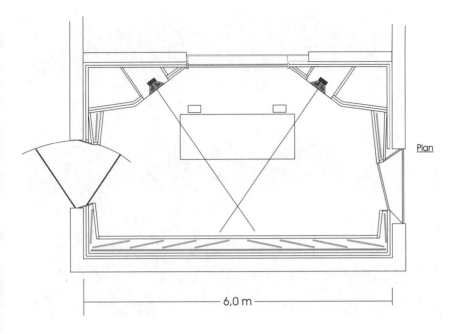

Plan

— 6,0 m —

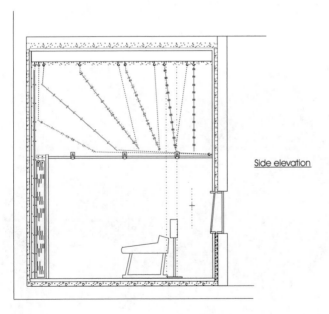

Side elevation

Figure 5.6 Studio Dobrolyot, St Petersburg, Russia: general plans of the structure

Figure 5.7 (a, b) Typical acoustic construction of a small 'Non-Environment' room. (a) The interior treatment of Studio Dobrolyot: the room as shown only awaits fabric surfaces, light and the installation of equipment

Figure 5.7 *cont.* (b) The finished room

are another variant. All the above are different in their composition and texture, but all perform a rather similar function. In the past, some varieties of roofing felt have been used, in fact by the BBC amongst others, but they tend to be highly flammable and inconsistent in their flexibility from batch to batch. The specifically manufactured acoustic products are all either non-flammable or self-extinguishing. Deadsheets will also be discussed, further, in Chapter 11.

The equipment in the above described rooms is carefully arranged so that the hard surfaces give additional life to the voices of people working in the rooms, but do not cause reflexions which could disturb the monitor response. If monitors are to be mounted flush in the front wall, the front walls should be made hard to function as effective baffle extensions, but if free-standing monitors are used, the front walls need to be acoustically absorbant, to prevent the omni-directional low frequency directivity of the loudspeakers from bouncing off the wall and returning to the listener in undesirable phase relationships with the direct signal, as previously shown in Figure 5.2.

These acoustic control techniques, described further in Chapter 11, are relatively new in their application to small spaces, as is the really significant shift towards producing such a great proportion of finished product in what were previously considered only to be writing and pre-production environments. However, even in these uses, when people have done such work in rooms which have been treated as described, the response is usually that the writing and programming becomes both easier and more enjoyable. The subsequent transfer of such work to the final production process also becomes more rewarding. When the monitoring conditions are good, it can be a surprise to many people just how much easier so many tasks become, but when the conditions are poor, the whole process tends to become something of a lottery.

References

1 Newell, Philip R., *Studio Monitoring Design*, Focal Press, Oxford, UK, p. xvii (1995)
2 Newell, Philip R., Holland, K.R. and Hidley ,T., 'Control Room Reverberation is Unwanted Noise', *Proceedings of the Institute of Acoustics*, *Reproduced Sound 10*, Vol. 16, Part 4, pp. 365–73 (1994)
3 Newell, Philip R. and Holland, K.R., 'Impulse Testing of Monitor Loudspeakers', *Proceedings of the Institute of Acoustics, Reproduced Sound 5*, Vol. 11, Part 7, pp. 269–75 (1989)
4 Clark, David, 'Precision Measurement of Loudspeaker Parameters', *Journal of the Audio Engineering Society*, Vol. 45, No. 3, pp. 129–41 (1997)
5 Colloms, Martin, *High Performance Loudspeakers*, 5th edn, Chichester, UK, John Wiley & Sons (1997)
6 Duncan, Ben, *High Performance Audio Power Amplifiers*, Newnes, Oxford, UK (1996)

Bibliography

Borwick, John, *Loudspeaker and Headphone Handbook*, 2nd edn, Focal Press, Oxford, UK, (1994)

Colloms, Martin, *High Performance Loudspeakers*, 5th edn, John Wiley & Sons, Chichester, UK, (1997)
Howard, David M. and Angus, James, *Acoustics and Psycho-acoustics*, Focal Press, Oxford, UK (1996)

Chapter 6

Modular digital multitrack recorders

One of the driving forces behind the rapid evolution of the project studios in the 1990s has been the advent of the modular digital multitrack tape recorders (MDMs) such as the ADATs and DA88s. They have offered a huge step forward in terms of achievable sound quality in their price range, but they have not been without their problems. In relatively simple use they have been a great boon, but where complex arrangements of these have been installed in high pressure working situations, they have not always lived up to the 'professional' label which some of their manufacturers have attached to them.

It should be accepted from the outset that machines using VHS transports, and which are built to a very competitive price, can hardly be considered to be truly professional, but that is not what past advertising has lead us to believe. To many people, such machines *are* perceived to be professional, but there is more to the 'professional' label than merely the performance capabilities.

My first encounter with ADAT machines, in April 1993, was a total accident. I had just completed the construction of a seven room recording studio for Regiespectaculo in Amadora, Portugal, and the first work was to be a mix for TV from a live concert of one of Portugal's best selling bands. I was asked along on the first day, just to see how things went and to help deal with any queries. After several hours, the engineers booked for the session still seemed to be totally engrossed in the wonders of the Amek 'Supertrue' automation system and the Alesis BRC 'Big Remote Control' for the three ADATs which they were using. Partly from boredom, partly from old instincts, and partly because I found myself very much liking the music, I began pushing faders about. Once some members of the band learned more of my 'track record' on live recordings, and hearing a style of mixing which they had not previously considered, the above arrangement of personnel remained for the duration of the mixes.

My initial reaction to the ADATs themselves was that I was hearing what was recorded, free of any noticeable noise, distortion or coloration, and I very soon began to appreciate the flexibility of the 3-machine, 24-track format. For example, all 24 tracks had been used, and when the possibility of replacing a badly out of tune guitar was being discussed, it was easy both to try it, *and* do it, without risking the masters. The method was simply to use two of the ADAT machines to copy the tape carrying the 8 tracks which included the offending guitar, put the master to one side, re-load the machines with the other two masters plus the digitally 'cloned' copy, then overdub the guitar. If it works, continue with that set of tapes, as the digital to digital copy is no different sonically to the master; and if the overdub does *not* work, then simply go back to the master, with the 'spare' copy reusable for other purposes. By the end of the sessions, I was very impressed by the whole concept of the ADATs.

6.1 A live report

As a result of the previous mixing, I was asked to organise the recording of a large outdoor concert in June 1993 at Alvalade, the stadium of Sporting Lisbon FC. There were to be six bands playing for 8 hours, with complicated instrumentation line-ups and mostly only 10 minute changeovers, mainly to prevent the 40,000+ crowd from becoming restless on a very hot afternoon and night. Fire hoses were frequently trained on the crowd to prevent people from fainting in the 30°C heat. Two of the bands were to be recorded for national television, with the other four being videoed by a commercial company. We decided to use six ADAT machines, five of them synchronised as a 40-track recorder, which also carried TV and video time codes for ease of editing. The sixth machine was used as a back-up, but was also able to run 'wild' in the event of other problems. It could carry audience tracks to cover the applause on changeovers when the other five machines were stopped for tape changes, every 40 minutes or so. If necessary, the 'wild' machine could later be re-synchronised using its own internal clock and a time code offset on the BRC. Again, the system was proving super-flexible. The only drawback was that this whole set-up was going to use over 120 tape cassettes, including rehearsals, and that meant an awful lot of pre-formatting.

6.1.1 Formatting problems

As it was not possible to rent the extra machines until the day of the concert, only the three machines from the studio were available. To format 120 tapes meant 40 runs of 40 minutes, formatting three tapes at a time, one on each machine. With re-wind and loading time, this was going to take over 30 hours of continuous formatting, and when the tapes finally arrived from the suppliers, there were just two days to go before rehearsal day. Going by my own experiences, and also from speaking to other people, it had been known for tapes to drop out on the first run, just stopping inexplicably. Once formatted, however, this rarely seemed to occur. What was more, on the machines which we were using, with the old 3.01 software, they would sometimes stubbornly refuse to begin to format, taking up to 10 or 15 attempts to lock into 'format' mode, and needing to be re-wound after each false start. We found that this could be overcome by placing a previously formatted tape in the master machine, then commanding the other machines to format, which they would do instantly. After ten seconds of formatting, the machines were stopped, the 'master' removed and replaced by one of the ten second formatted slaves, then, the empty slave machine was loaded with a blank tape. When all machines were re-wound, re-formatting of all of them would commence without trouble. This was typical of many strange characteristics that would become apparent as the months and years passed by.

It *is* possible to format as you record if the tape will be run continuously from beginning to end, as is the case with such a live recording, but given the inconsistencies of the initial formatting process, it was decided that it would be absolutely necessary to pre-format all the tapes before the show. At least then we could be reasonably sure that all the tapes ran from beginning to end without problem, and would hopefully start promptly.

I decided to take the three machines and all of the tapes back to my hotel, when I could spend a couple of days sunbathing on the balcony, with a trip to the shade every 40 minutes or so to remove the formatted tapes and re-load. All went very well until the evening of the first day of formatting, when one machine suddenly indicated 'Protect' on a tape where clearly the break-out 'write protect' tab was still in place. Come what may, no matter what tape was loaded into the machine, the machine stubbornly displayed 'Protect' [Prot]. Given the late hour, 9pm local, I soon calculated that there was not enough time left to format what remained on only two machines, so I decided to tele-phone the manufacturer in Los Angeles. The first problem was that the manual only had one of those American 'toll free' numbers; totally useless when dialling from anywhere outside the USA, and especially from countries with only numbers and no letters on the phones. More by luck than anything, I found a magazine containing a review of the ADATs, which fortunately con-tained a telephone number.

6.1.2 Unprofessional back-up

Usually, American companies are ultra-helpful, so I was quite shocked to be confronted by a person who sounded like a stuck gramophone record repeat-ing 'Sorry, but it is not company policy to divulge such information directly to users', in reply to my request for guidance as to what would be causing my 'write protect' problem. Out of desperation, I did something which I truly hate doing, and I told them that they would be reading a world-wide put down in the press if they refused to budge on this. Even after a third request for me to 'hold' whilst discussions took place in Los Angeles, I was still told that company policy was final. I had explained that over 100 technicians were working on this concert, coming from many countries to work on the pro-duction of this very expensive event, and eventually I was put in contact with somebody associated with the manufacturer who, privately told me the answer to my problem. Within 10 minutes, I had all three machines running again. I also learned that the problem was a well-known fault to the manu-facturer, and received other gems of information which were crucial to my five machine synchronised set, yet which appeared nowhere in the manual; such as when using machines with a mixture of software, always use the *oldest* software as the master, as the older versions may not be able to follow the newer software's commands. Such information might seem to be obvious now, but when the machines were in their infancy, we were all groping in the dark.

The manufacturer's stock answer to my queries had been a very 'official', 'Consult your local dealer,' but at 9 o'clock in the evening, the dealer was closed. We did not have 12 hours to lose, as even with all three machines going by the next lunchtime, there would simply not have been enough time left to format all the tapes. On top of this, in certain far-flung places, consulting your local dealer is about as informative as discussing the problem with your local astrologer.

My number one complaint became centred around the word 'Professional' which was proudly printed on the front of each machine. Usually, the meaning of the word 'Professional' does not necessarily imply superior quality of performance, but more that the device can be expected to work consistently and reliably over many years, and under arduous circumstances. Somewhat more importantly, it also suggests that excellent back-up is provided, with helpful, knowledgeable people ready to give whatever assistance they can, either in person or at the end of a telephone line.

Irritated by much of this, I wrote a letter to *Studio Sound* magazine, who published it in October 1993. Far from eliciting a constructive reply from the manufacturer, one main importer called the magazine threatening to withdraw all advertising for daring to publish such a critical letter. Another importer (well the business brain of the company at least), in another country, was heard to say that I should not say the things that I was saying because it could affect their sales.

It certainly could; and so it should if they were selling the machines as professional but not giving appropriate back-up. One question which therefore carries considerable weight is 'Can one afford to risk using equipment which does *not* have a helpful back-up service available on an unrepeatable and expensive recording event?' The question would seem to be not so much whether the ADAT per se is 'professional', but more whether the manufacturers adopt professional attitudes. However, to be fair to the ADATs, I know many owners of other types of MDM who are left in similarly helpless situations when crises arise.

Anyhow, notwithstanding all of this, the Alvalade recordings took place as scheduled, and were eventually used in three TV shows, three commercial videos, and at least one CD. The recording went very well, with only two minor problems. The first was that, at least with the software which we were using, the five machines could take anywhere from 4 to 17 seconds to lock up. In actual fact, however, we only lost the introductory vocal off one song out of the whole day's recording, and that would have been no problem to replace if it had been needed. The second problem cannot really be charged to the manufacturer, as we had two Maxell tapes drop out and seemingly stop. The manufacturer clearly said at the time to use only recommended tapes, but at four days notice, in the whole of Portugal there were not the 120 tapes which we needed. Under the circumstances, we bought all that we could, and had to use a few SVHS cassettes of other manufacture. Interestingly, the two tapes which dropped out and stopped worked perfectly when re-formatted and repeatedly tested over the course of the next week. The problem was never found, but it seemed to be something in the initial formatting, as though the formatting process had dropped out but the machines somehow continued, at the time, to run to the end.

The recordings themselves were very good. The excellent signal to noise ratio of the ADATs allowed a good safety margin of headroom, which is essential when recording only partially rehearsed events. Where levels were lower than intended, the noise floor was still low enough to keep everything perfectly usable. Over the course of the following six months, I was using ADATs for recording many different line-ups, from rock bands to vocal ensembles and string sections, and I certainly have no complaints about the sound quality of any of the recordings.

6.2 A test of quality and a choice of levels

Shortly before Christmas 1993, I was asked to record three concerts at the Centro Cultural de Belêm, in Lisbon. Aside from a few guest participants, the concert was essentially by two renowned pianists, one each from the worlds of classical and jazz music, performing a varied repertoire from both domains. The pianos were a pair of huge Steinway concert grands, beautifully tuned and with penetrating attack. The microphones used were mainly pairs of Schoeps CMC56 'Colette' series, in stereo pairs over each piano, and a separated pair on the audience. Single AKG C451s were also used, underneath the piano soundboard and over the centre of the audience for 'fill-ins' when and where required. We had a purpose-built television sound control room in which to set up the recording equipment, and an old but very clear pair of KLH Model 5 loudspeakers on which to listen, driven by a Crown PSA2.

The mixing desks used were Mackie 1604s, and here came the first dilemma. The Mackie mixers have a very clear signal path, and outputs optimised for OVU (1.23 V), a level which when fed into the EDAC inputs on the ADATs, would put a level on tape of 18 dB below peak recording level. It is well known that to digitise a signal with the maximum number of bits, and hence maximum low level signal resolution, the peak recording level should be '½dB' below the maximum recording level of the system. To do this, the Mackies would have to drive 18 dB over the +4 dBm (1.23 V) nominal output level, or +22 dBm. Whilst this is within the maximum output capabilities of the Mackies, it is very close to overload, and hence there is a question as to whether a line output amplifier stage would be at its most sonically pure when running so close to the limit. It was decided to go into the −10 dBV inputs on the ADATs, still feeding them with the '+4' Mackie outputs, and driving the Mackies to record a peak level close to the limit on the ADATs; hoping that the '−10' input circuits would not suffer undue sonic degradation when receiving an input level 18 dB over nominal.

Why manufacturers of much digital recording equipment do not provide input and output gain controls, as they do on most analogue machines, I do not know. As the above problem shows, it is not easy to set optimum recording levels on tape from a standard input level. I realise that on desks with VU metering, peaks can greatly exceed the nominal levels, and working 'up to 0' on the mixing desk meters will probably require a headroom reserve on the tape. Because digital tape overloads grossly, without the soft limiting and compression of analogue tape, a chance sudden overload is much more likely to be catastrophic; hence the safety headroom requirement provided by the ADAT manufacturers. On the other hand, if the mixing desk is fitted with fast, peak reading meters, then peaks can be monitored and controlled from the mixing desk. But, ... few mixing desk meters read +22 dBm, so unless extra gain can be provided at the record input of the machine, when driving the tape to full level, the mixing desk meters would tend to be more or less permanently off the scale, which renders such readings useless. The subject will be dealt with, at greater length, in the following chapter, but suffice to say here that lack of recording level control can be very frustrating.

Anyhow, the recordings in Belem were set up as described, and the signals monitored *through* the ADATs. One Mackie was used for the recording channels, and the other was connected to the machine outputs as a monitor desk. In this way, with the machines on standby and reading line input, the whole signal chain could be monitored without the risk of missing a wrong or bad connection. Equally, when the recording had been made and the machines were put into play, an immediate assessment could be made of the recording. During the final rehearsal, we tried a few short test recordings to verify that all was working, and all appeared to be satisfactory.

After the first night's concert, a show of almost 2 hours, I quickly ran the tape back to what I felt were some of the most exciting parts of the performance, then set the machine into play. During the recording, I had been very impressed by the dynamics, clarity and full body of the sound, and when watching from the wings of the stage, I felt that I was definitely listening to the selfsame pianos that I was recording. The monitoring level in the control room was quite life-like, and the mood of the music was 'travelling' well. On first playback of the recordings, after almost 2 hours of listening to the show, there *was* a slight loss of transparency and physical depth in the reproduction, but the degree of this was quite impressively small. Indeed after 10 or 15 seconds of the playback, my mind began to lose track of the differences. I have been involved in classical piano recording since 1970, when I worked for Pye Records, and compared with the usual quality loss on playback from analogue machines, the loss from the ADATs was minimal. Considering the cost of the ADAT, and that each one contains 8 D to A and 8 A to D converters, *and* considering that a single *pair* of audiophile convertors can cost around the same price as a whole ADAT machine, then the cost to performance ratio of the ADAT becomes quite remarkable.

The real acid test came a few weeks later, when we were listening to the recordings for mixdown and editing. The mixdown engineer was Branko Neskov, a Yugoslav who is more usually to be found working on film soundtracks. Together with Branko and myself were the two pianists (one of whom played two concerts in Lisbon as soloist with the London Symphony Orchestra under Sir George Solti). The pianists, Pedro Burmester and Mario Laginha, sat at the mixing console after Branko and I had discussed and set up the basic listening balance. No equalisation had been used on the recording, and the small amount used on mixdown was mainly to help to remove some fan noise, caused when somebody turned on the auditorium ventilation fans during the performance, a problem corrected on the second and third days of recording.

In the studio, nobody commentated about the sound quality, but very soon, the pianists were 'playing' the mixing console and were very animatedly and excitedly moving. The suggestion to Branko and I was that they were hearing something very close to what they heard on stage whilst sat at their respective pianos. By April, and some 300 hard disc edits later, we had an album (a composite from three performances and four rehearsals). The musicians had not once been critical of their beloved piano sounds. Given the previous successful recordings which I had made with ADATs of rock bands, string sections, vocal sections and electronic music, these piano recordings just about set the seal of approval, in my opinion, in terms of the recorded sound quality from the ADATs.

6.3 Head cleaning and other problems

During the Belém recordings there *were* problems, but we achieved what we set out to do. The first problem was before the second performance. The day after these piano recordings we were due to begin recording a rock/pop band for three nights. This would be another 40-track, five-machine set up as at Alvalade during the previous June. We calculated that to run five machines for a two-hour concert on each of the three nights, plus a spare set of tapes, would need 65 SVHS cassettes. Again, the time consuming formatting could not reliably be expected to be done on each recording day, so we decided that before going to dinner, we would set three cassettes formatting, and try to format as many as possible whilst still set-up for the piano recording. To our horror, when we returned from dinner and tried to play the tapes, the machines were flashing their drop-out lights, and on one machine, the light was almost permanently lit. Again, the recommended tapes were not available in Portugal in the quantities which we needed. The importers of the machines had recommended a TDK tape, but their mechanisms were *very* noisy and at least 10 per cent of them were clogging the heads.

Under circumstances like this, the 'Do not clean the heads' and 'never with any liquid' instructions in the MDM manual are totally out of place. We had TV trucks outside and a concert to record one hour later. A quick trip to the chemist produced a bottle of alcohol, and with the aid of a soft cloth (relatively lint free) the heads were cleaned. This brings up again the problem of the manufacturers calling these machines professional, yet instructing users to return them to the importers each time that there is a problem. Clearly in the circumstances above, to send the machines from Lisbon to Oporto, some 300 km away, for the heads to be cleaned was not on; we had a concert to record.

The second problem at Belém was a 'one off'. At the beginning of the second half of one show, the master of the two machines used for the main recording (there was a third running wild to cover the audience and TV time code during tape changes) was put into record. The slave began running, and although the master showed its record and play lights lit, the tape timer remained at '0000'. The machines were re-wound twice and started again, only for the same thing to happen. The tapes were then ejected and re-loaded ... same result. A new pair of tapes was then loaded and the machines went into record perfectly. Subsequently, the suspect pair of tapes were used to record a rehearsal with absolutely no problem whatsoever. I have never seen that problem again. The problem did cost us the recording of 'Scaramouche' but fortunately, that was needed for neither the TV programme nor the CD, but even if it had been needed, the 'wild' machine was already running and covering the most important microphones.

6.3.1 Error messages and professional needs

Back to the subject of the 'professional' claims for the machines; let me just relate a couple of sets of circumstances which highlight my points. On the mixdown of the above recordings, we were using a pair of machines which were approaching 1000 hours of use. There were times when those infuriating Er2, Er5, Er7 error messages would flash up on the displays. Although by dubious means I have gained knowledge of what those messages mean, to the majority of the users, they are totally unknown, and distributors are prevented by their contracts with some manufacturers from disclosing their meanings to third parties. This is a sure sign of a 'domestic' marketing strategy. I know one very respected engineer who says, 'How can I feel confident in using a machine in which I do not know how serious a fault may be? In other words, should I stop the session quickly when I get an error message, or is it safe to carry on and take a risk? Will my tapes be ruined if I continue? Will the recordings be usable long term?' On a different recording session, we had one situation on a Friday night when repeated error messages were being noticed. What was more, the BRC display kept indicating 'Error on machine 2' without seemingly any way of cancelling the message once it had been noticed. We were trying to do drop-ins and relying on the BRC counter for cues, but time and time again, ten seconds before drop in on a repetitive section, 'Error on machine 2' would flash up, replacing the time display. We were thinking 'OK, there is an error on machine 2, we know all about it, it has told us twenty times, ... but what is it? What can we do about it?' What is more, the error message made it impossible to see the time at the crucial drop-in point.

One of the musicians was off on a tour of Germany the following day, so this was a last chance to do the recording. To send the machine back to the importers could not be done till they re-opened on Monday; and then the 600 km round trip would probably mean Wednesday before we had the machines back. Five days without machines would not keep a professional studio professional for very long. Despite having skilled technical staff around, the lack of information with regard to error messages rendered them to be of little help.

6.3.2 Philosophy

The 'write protect' problem referred to at the beginning of this article was a typical piece of unskilled user protection that caused problems. There had been some faulty micro switch levers in the field, which had led machines to believe that the write protect tab had been removed from the cassettes when in fact they had not. This micro-switch totally disables the record or format functions of the machines, though I can think of no truly professional reel to reel machines with any similar system. On professional machines, pressing the play and record button puts the machine into record unless the 'channel record' switches have been deactivated. There is no 'record protect' tab on the reel flanges. This seemingly domestic safety feature of the ADATs creates one extra source of professional problems with little, if any, benefit to the professional user.

MDM manufacturers must face something of a dilemma here, because for the very remarkable price of the machines, I suppose that one is not really justified in *expecting* a professional machine with professional back-up, but if that is the case, then why market them as what they are not? The machines are truly wonderful for the price, and in one incident in November 1993, I saw a tape which had been chewed up in a machine (*not* a manufacturer recommended tape I may add) copied as best it could be, then be entirely re-built from a similar chorus section using another machine and a BRC. Coincidentally in the same week, I saw a chewed up tape from a fully professional $100,000 24-track machine become a total write-off; though it must be said that this is a very rare event on such machines. Perhaps professional users would pay considerably more if manufacturers would produce truly professional machines, with input and output level controls, no unskilled user protection, and for which they could provide a professional customer back-up, *including* precise details of any error signals. Even at two or three times the price, that would still be a cheap machine given its capabilities. (Note: As this book is going to press, such machines are just emerging, but it is too early to judge the market reaction.)

On balance, what I actually 'lost' from five live concerts whilst using the ADATs is still absolutely minimal. I remember in the late 1970s, whilst I was part of the team recording (or trying to) four concerts of Little Feat at the Rainbow in London for the 'Waiting for Columbus' album, a pair of Ampex 2 inch 24-track machines absolutely refused to run consistently on almost all of three out of the four concerts. I ended up nursing the machines through the fourth day, which luckily was the only one where the band played well, but oh to have had the ADATs then. What a luxury it would have been. The big difference, though, was that a phone call brought engineers directly from Ampex to help us when problems arose. Had the ADATs failed, we could have been on our own.

6.4 104 tracks

In 1995 things were really pushed to their limits. Figure 6.1 shows a set-up, again at Alvalade, using 13 synchronised ADATs in a 104-track configuration. The job here was to record 19 bands in 5 hours, with a minimum of changeover time. From a concert without serious rehearsals, we recorded 37 songs out of 43. Synchronisation problems caused the loss of at least part of the remaining 6 songs, but nevertheless, the concert organisers considered this to be a success. If pushed, four out of the six 'lost' songs could have been salvaged. This concert was recorded at short notice, and the ADATs were a collection of machines from many different sources, and with quite a mixture of software.

The recording was made in the summer, and the temperature was over 30°C. Test runs prior to the concert revealed a few problems of various sorts, but it was found that the machines were running very hot. The application of ventilation fans and a portable air conditioning unit greatly reduced the problems, which suggested a degree of heat sensitivity which had been suspected in the studios, some months before. The infuriating thing, once again though, was that the error indications which the machines were giving would not be explained by either the manufacturers, or the importers.

During this event, the stage was set up with two complete sets of basic instruments, so that one band could be getting ready to begin playing whilst another was still performing. In addition to the basic instruments were all of the specific instruments for the individual bands, which in all meant about

Figure 6.1 104 tracks of ADAT. A temporary set-up for the recording of nineteen groups in the football stadium of 'Sporting Lisbon' in Portugal, in 1995. One Yamaha and four Mackie consoles were used, operated by three engineers

ninety lines from the stage to the recording consoles (five Mackies and a Yamaha [used as mic amplifiers only]). On top of this were audience microphones, video soundtracks, front of house mix, and time code. Two or three days before the recordings, a general request was put out for ADATs, and eventually, with the last two arriving only 30 minutes before the show began, 13 were assembled into a single 104-track digital recorder, controlled by an Alesis BRC.

After several companies had rejected the concept of recording such an event, which once again included many famous artistes, Regiestudios invited me to a feasibility meeting. We were given the go-ahead only eight days before the concert, but at that stage, the studio only had its three ADAT machines and around twenty tapes. Respecting what I have said previously about pre-formatting, we decided that we would need to format around 110 tapes, and this time, the three machines were only available at the week-end, being in use in the studios the remainder of the time. The go-ahead was given on the Wednesday with the concert itself taking place on the Thursday of the following week. Between the go-ahead on the Wednesday, and the availability of the machines for formatting at midnight on Friday, 110 tapes needed to be located for formatting.

Once again, there was nothing even approaching that number of recommended tapes in the dealerships of the whole country, and by the time that this was realised on the Thursday, there was no longer time to import them. The decision was taken to go ahead with Sony VXSE 180s which we had used very successfully before. Around midnight on Friday, the three machines and the tapes were brought to my hotel, and the formatting continued all through the days, and most of the nights; finally being completed about 9am on the Monday morning. The problem from the previous year, of some machines being reluctant to begin formatting an unformatted tape, still existed. Once again, the problem was solved by 'seeding' the master machine with a pre-formatted tape. With the newly installed 3.04 software, the problem did not seem as bad as before, but nonetheless, a problem it still was.

After about 36 hours of almost continuous use, one machine decided to eat a tape. This *could* have been an overheating problem, as it was the machine in the centre of the stack, and the afternoon temperature in Lisbon was over 30°C in the shade. It was found that the take-up hub of the cassette was not pulling the tape back inside, so the machine was sent back the next day to the distributors, where it was found that a new drive belt was needed. Another machine was borrowed for the next two days of recording in the studio, and the faulty machine was returned in time to go to Alvalade on the Wednesday, the set-up day before the day of the show.

When the 13 machines were finally assembled, they had a mix of 3.03, 3.04 and 3.06 software. They were all driven from a BRC, but nonetheless, in the event of the machines having to be controlled from the No. 1 machine for any reason, they were daisy chained in rising order of their software numbers, machine 1 being 3.03 and machine 13 being 3.06. In multiple stacks, remember, the oldest software should always be at the head of the chain.

We were to receive little or no co-operation from the stage in terms of providing announcements or other pauses in the music in order to give us the time to change tapes, so to try to cover the changes as best we could, a Fostex E16 was run during the changes, carrying the audience microphones, the lead vocal microphones, and the front of house mix. It was re-patched each time to carry whichever instruments were to play first: plus of course, time code. A time code DAT machine was also used to record the front of house mix.

In practice, with four people changing tapes, two removing and two loading, we could change all 13 machines in about 30 seconds, including the ejecting and re-synchronising time. The fear was that in the pressure of the concert, we would end up with a pile of 26 cassettes on the floor. Luckily, this did not happen, but three or four of the eight changes were not as smooth as hoped for. Of the begged and borrowed machines, three showed problems. One machine, in about 30 per cent of cases, would eject a tape within a couple of seconds of it being loaded, for no apparent reason. Usually though, the second loading was accepted without problem. Another machine was reputed to have tracking problems, but these disappeared after we lifted the lid and cleaned the whole tape path. A third machine infuriatingly intermittently displayed the 'PROT' message when attempting to put it into record. I say again, those amateurish 'record protect' tab sensing switches have no place on a supposedly professional machine, where they are nothing but a liability. I would recommend that any person still seriously using the earlier ADAT machines should cut the pink wire going to the vertically mounted tab-sensing switch.

6.4.1 The concert

The way that the recorded tracks were laid out, some of the machines carried nothing when certain bands were playing, which was to be useful later in the show when some machines showed erratic behaviour. At the very beginning of the show, two machines failed to go into record, one displaying the 'PROT' signal, and the other for no apparent reason. All attempts to get the rogue machines to record were unsuccessful, so the whole stack was re-zeroed with the 'locate zero' button, and on the second start, after ejecting and re-loading the tapes, all machines synchronised. Luckily we lost nothing important. On the third change we had a very tight changeover, with drums more or less continuous through the change. By this stage, one machine had become free, as the instruments allocated to it were for use with bands performing in the first hour. I decided, possibly rashly, to disconnect the last machine in the chain, so that it could be run unsynchronised from the main stack, with the important drum tracks parallel-fed to it during the tape change. The machine could then be re-synchronised later from its internal clock.

On the change, the machines failed to record, possibly, and here I am guessing, because the BRC was still 'looking' for machine 13, which nobody had 'told' it had been disconnected. The only recourse here was to switch off the BRC and switch on again after a suitable pause. On the re-identification of the machines, we ended up with three machine 2s and two machine 7s, so we had no option but to switch off everything, pause, then turn it all on again in sequence, BRC last. This time all was OK, but one song was lost in the process.

In two of the subsequent changes, two machines refused to record, but by this time, more machines had been freed of their recording burden, so the inputs to the non-recording machines were quickly re-plugged into the redundant machines. Any machine not recording was allowed to continue running in play, as to have 'missing' tapes from the pre-arranged order would have led to worry and confusion when the time came to sort them out. In the struggle to keep everything going, making detailed notes was not possible, so when the inputs were changed, so were the machine and tape labels. In other words, if the inputs from machine 7 were re-plugged into machine 3, then the sticky labels were exchanged on the machines, and the pre-numbered tapes were fed into the correspondingly numbered machines, irrespective of their new positions in the stack. On the last two tape runs, the 12 synchronised machines, plus the 'free' machine, all ran and recorded perfectly, though by now, only about seven of them were carrying any necessary signals.

I had told the organisers before the show that without co-operation from the stage we could lose up to ten songs out of the forty-three or so to be performed. The prime reason for the recording was to make a special television feature programme, which would probably be around 2 hours long at the maximum. The organisers said that the live show was the main event, and if we lost ten songs, then they would still have over 30, which was sufficient for the TV show, so to do our best. In reality, I believe that we lost six, and only two beyond repair, so by the end of the night, there were no complaints.

I must admit to being mentally and physically exhausted the next day, and when somebody called to ask how the concert went, my initial reaction was 'What concert?' ... I was still waiting for it. I later began to put together a picture of events on stage (we had been surrounded by seven television and video monitors) and from what people had said to me, but although I remember well the recording, I remembered little of the performance or the show. Clearly, it had not been a stress-free event.

6.4.2 The post-mortem

Considering the 11th hour arrival of some of the machines and the consequent inability to check the whole system prior to the show, to lose only six songs was not bad going. At the end of the night, we handed thirteen working ADATs back to their owners, and had over 100 cassettes recorded, plus three ½ inch tapes from the Fostex E16 and 3 DAT cassettes. The 'building block' concept of the ADATs is a very, very flexible and affordable system. I suppose that we *could* have used a pair of synchronised Sony 3348 48-track digital machines, plus a 3324A or S, but where would we have found them at short notice? The cost and difficulty of transporting them could also have been prohibitive. Basically, ADATs made this recording possible. The nearest option on offer was to use a pair of synchronised 24-track analogue machines, but the 46 tracks available would have necessitated too much pre-mixing of a very unpredictable event, and any required repairs to the recordings would have been much more difficult and time consuming.

The mixdowns were ultimately not done from thirteen machines, but from three. Those same three machines were used to digitally bounce the live tapes into a more condensed form, then sub-mixes of sections or drum kits could be done in the studio, under good mixing and monitoring conditions, and with time to make the appropriate decisions about sounds and balances. If any of the sub-mixes were ultimately deemed unsuitable in the final mix, they could easily be re-made from the original masters. For this type of work, in my opinion, the modular digital 8-track formats really have no equal. The sub-mixes and final mixes took about three weeks of studio time, and were ultimately used for *two* 75 minute television programmes. Approaches were made for permission to use some of the recordings for live tracks for CD release, so obviously, the results worked out very well.

During this recording, the importers were *very* co-operative, with the owners coming down in person and proffering spare machines, spare software, spare cables and whatever else was required. Although they were somewhat upset by my previous criticisms of the system in the press, I think that once they witnessed the stress of such recordings, they realised that my previous statements had not been unjustified. They had been the result of the frustration of seeing a potentially fantastic concept spoiled by lack of professional back-up, going all the way back to the manufacturers themselves. But, I state again most clearly, the modular digital 8-track format made this recording a practicable reality.

6.5 Comparisons in sound quality and typical usage

Since those early days in Lisbon, I have built many studios which have used MDMs in some very professional operations. In fact, they have used the Alesis ADATs, the Fostex ADATs, Tascam DA88s, and their Sony counterparts. I have also built several studios using Sony DASH (Digital Audio, Stationary Head) machines. I honestly cannot say that I have detected any significant differences in the sound quality of the various formats or makes. In fact, in the April 1997 *Studio Sound*, there was a lengthy report of a set of listening tests which showed that a panel of expert listeners had very great difficulty in perceiving significant differences between a range of digital machines, especially when using the digital inputs and outputs. (Figure 6.2 shows a Sony 3324A, 24-track DASH machine, in the machine room adjoining the control room shown in Figure 2.1.)

In these tests, some degradation of signal quality was noticed when using the D to A converters of the individual machines when compared to a very expensive Apogee D to A, but in no case was the difference reported to be very great. From my own experiences, I did not find the results of these tests in any way surprising, but it still does surprise me how, when first class D to A converters are so expensive, the inexpensive, modular 8-track machines can supply 8 built-in devices of such good quality when the entire machines cost so relatively little. The overall sonic quality of the various machines has always seemed excellent to me, so I have been surprised at the times when people, especially in mid-priced studios, have been critical of their sound.

In an industry when fact and fiction are so often blurred, carefully constructed presentations such as the one in *Studio Sound* provide an excellent anchor to reality. So, why are there so many people going around saying that certain digital recording systems are poor? This type of comment is prevalent in many project studios. Well, I believe that almost all of these comments are because they are blaming the recording machines for the sound of entire recording chains. The reality is that very few people make recordings on modular 8-track machines (relative to the size of the market, that is) using sets of $3000 microphones and $5000 FM Acoustics 'ClassAmp' microphone preamplifiers. Neither do they often monitor the proceedings through first class monitor systems. The truth is that the more professionally orientated

Figure 6.2 The Sony 3324A digital tape recorder, in its air-conditioned alcove. This alcove forms the machine room of the control room shown in Figure 2.1

machines, such as the Sony 3324S, are more frequently associated with systems using very expensive, 'high-end' microphones and mixing consoles. Very few studios buy Focusrite or Neve mixing consoles then record on ADATs. In the majority of cases, therefore, the recording machines tend to be mated with other equipment of a corresponding price level. As with the old computer business adage "rubbish in equals rubbish out", so the recording machine should not be blamed for what it reproduces faithfully if the source signal is compromised.

Two other aspects of the project studio world can also blight the less expensive machines. Firstly, people of lesser levels of training and experience tend to be associated with a greater proportion of the recordings made on the less expensive machines than with the top of the range machines. These are the people who will be most likely to use +4 dBV balanced inputs, come what may, believing them to be 'more professional' even in cases when the

−10 dBV inputs would be more appropriate for optimal interfacing with other equipment. Secondly, the level of musicianship and production associated with the greater proportion of these machines may be of a lower standard than the greater proportion of work utilising Sony 3348s and the like. Thus, in general, the higher priced machines tend to benefit from receiving a higher quality input. It is thus no surprise that what they are usually heard to reproduce may be more spectacular than the output from the average modular 8-track machine. In my own experience, sonically speaking, the modular 8-track machines leave very little to be desired, especially when using their digital inputs and outputs.

If only such could be the case with the mechanical side of their operation. Most of the studios which I have built using such machines have reported frustration with their day to day operation. For those with studios in the same cities as main agents for the machines, the maintenance is not too problematical, but when a city is 500 km from the nearest service centre it can become very difficult to provide a professional service. In fact, in some countries, I have known of main agents whose servicing ability is so poor that machines have been returned with one fault fixed and two more created. It is not uncommon for machines to be returned from servicing with head misalignments such that they would no longer interchange with the other machines in the studio. Such things make a mockery of professionalism. There also seems to be some doubt about the quality of tape available in many places. If studios have access to specially graded tape, such as is available from certain specialised recording materials suppliers, then they may not have problems with some error indications, but there seems to be evidence emerging, even for DAT tapes, that some parts of the world, and some less favoured suppliers, are not receiving the best quality tapes from the manufacturers, and this can lead to reliability problems.

The question of routine maintenance is also a hot topic. John Watkinson, one of the world's leading authorities on digital recording systems, is most emphatic about the fact that a tape machine is a tape machine. Whether it be digital or analogue, domestic or professional, it needs regular cleaning and alignment whenever it seems to be obvious that it is required, and whether this coincides with a regular service interval, or not. If tapes shed any oxide, then the heads may need cleaning. For any self-respecting machine which purports to be for professional use, this must be able to be carried out by studio staff, or the situation is farcical. Recently, though, Studer have begun producing serious modular 8-track machines, with proper service manuals, so perhaps the less professional machines will in many cases be relegated to home use. The point is that if opening a machine for urgent, necessary service invalidates a warranty, and *not* opening the machine incurs the cost of a lost session in a serious set of circumstances, then the machine cannot reasonably be called 'professional'. Serious problems at some of the live events already described could cause a much greater loss of money than the cost of buying a new machine, not to mention the loss of reputation.

So, if the warranty loss becomes nonsensical in the context of the many working circumstances, it only makes sense to ignore all the 'Do nots' of the manufacturers, and treat the machines like professional machines should be treated – with regular (daily, if necessary) maintenance. Statutory rights should not be affected, as many machines are aggressively marketed as being professional. However, if the machine causes significant loss, there may be recourse to damages. Most countries have laws relating to trade descriptions, and misrepresentations. If a device is marketed as professional, and its realistic performance, over a period of time, clearly fails to meet necessary professional standards, then legal actions may well be possible. I mentioned earlier that, coincidentally, on one day when I witnessed a tape being 'eaten' by an ADAT, a colleague lost a tape on a Sony 3324A. Indeed, that was the case. Nevertheless, the 3324A in question has now been running, at a very high work rate, for more than seven years, in a studio with relatively little maintenance back-up. In fact, the Sony DASH machines have deservedly earned themselves a truly remarkable record of reliability.

6.5.1 Financial realities

About two years ago, I rebuilt a studio in which the owners decided to buy ADAT machines of Japanese manufacture. Since then, the number of failures in heavy daily use has been running at a very frustrating level, compounded by the problems of getting maintenance back-up at awkward times. The studio often runs 24 hours a day. The chief engineer recently calculated that over a ten year period it may be necessary to replace the three machines five times; say once every two years of heavy use. Couple to that the cost of maintenance, the cost of transport for sending the machines to and from repair stations, the expenses due to lost time, *and* all other general running costs, then the use of three ADATs would probably cost over 50 per cent more during ten years of use than using a Sony 3324S. What is more, the resale value of the 3324S would be much greater, even after ten years, than the fifth set of ADAT machines. The calculation already included the resale of the other four sets at the end of each two year period, plus the differences in bank interest between the two methods of purchase – the Sony in a lump sum, and the ADATs every two years.

6.6 Summary and observations

Figure 6.3 shows a studio using four Fostex ADAT machines. The studio is owned by Rui Veloso, the Portuguese singer/guitarist and was constructed in the outbuilding, shown to the right of his house in Figure 6.4. These are used largely for his own recordings, and he also has a single machine in his house, for use when songwriting. For people using the modular 8-track systems as single machines, the reliability seems to be much better. Most of the time wasting and many of the general problems with MDMs seem to relate to the synchronisation of multiple machines and the interchangeability of tapes between machines. Up to now, I have built around 20 studios using three or more modular 8-tracks. There has tended to be fashions of one machine being in vogue, and then another, but enough time has now elapsed to be able to form a general picture. None of my clients is entirely satisfied with the reliability of the machines, and all express frustration at the need to send these away each time that a problem occurs. During the same time, the owners of the studios which I have built using DASH machines have nothing but praise for them. The modular 8-tracks are excellent value for money for light-duty

Figure 6.3 Fostex ADATs. The control room of the private studio of Rui Veloso: 32 tracks of Fostex ADATs are visible, along with an early Soundcraft DC2000 mixing console and KRK monitors.

Figure 6.4 The outhouse building to the right-centre of the photograph contains the control room shown in Figure 6.3

use, and the quality in the recordings which they can produce is first rate. However, I would not choose to use them in heavy usage situations on expensive recordings, such as where many session musicians were involved.

The experiences described in this chapter will, hopefully, help to put into a realistic context some of the experiences which will have been shared by many modular 8-track users, and the suspicions of prospective users. There is little evidence that the later versions than the ones described in the reports in this chapter are significantly more trouble-free. For many people, there is a tendency, at times, to feel alone with their problems, but maybe, the evidence presented here will give them some comfort, and perhaps a little more confidence if they wish to take more control of their situations. When used in relatively light duty situations, I have no hesitation, whatsoever, in recommending any of the machines. In terms of sound quality, the differences between them are absolutely minimal; but this can only be verified by controlled tests which are fair to each machine. I have known people criticise certain machines when they have only heard them in one location, where perhaps they have been non-optimally interfaced, or have been connected to a less ideal signal chains. My reservations about the use of the modular 8-track systems mainly surface under conditions of heavy daily use, or on one-off recordings, where reliability is paramount. Only time will tell if the latest generation of significantly more expensive ($7000), heavier duty machines will show better long-term reliability. They certainly appear to be much more rugged than their predecessors. It should be remembered, though, that the really big profits in producing these machines comes from mass sales. Mass sales are very price conscious, so it is unreasonable to expect that the companies who are mass producing the machines will be too inclined to spend too much time worrying about the problems of the few serious users. They will go for the quality/price compromise which yields the greatest profit.

6.7 Geographical variability

In late 1997, reports began to be published about compatibility zones in the USA. Machines in the east coast regions were remaining compatible with other machines in the region, but would often cause problems if recorded tapes were sent to southern, or west coast studios. Inter-regional interchanges were generally becoming a problem. It was beginning to appear that regional distributors were developing their own alignment procedures, which was leading to inter-machine compatibility being restricted to specific geographical regions. It would appear that we still have a long way to go before the modular digital multitrack systems realise their full potential. Who knows, though, whether the introduction of new recording systems, coupled with the experiences of past frustration, will render these concepts to be obsolescent before they ever reach the hoped for state of development.

The lasting impression from six years of experience with the MDMs is of a seemingly endless stream of annoying and frustrating problems limiting the use of an exceptionally flexible and potentially remarkable recording concept. Now, in mid-1999, many studios seem to be changing from MDMs to hard-disc systems. Whether they are wise to do so, only time will tell, but most of the people to whom I have spoken cite the MDM reliability problems, and most definitely not their recording quality or functional flexibility, as the reason for the change.

Hardware-based recording systems need individual attention when things go wrong. Software-based systems can be updated by revisions over the Internet, so can be 'mass-repaired'. As the cost of physical maintenance is not too different for an ADAT or a 3348, the economics make less sense if a semi-professional hardware-based system needs frequent attention. As marketing demands make it necessary to release half-developed systems, these days, perhaps the software-based systems make more economic sense at the lower end of the market, because software repairs are relatively cheap to distribute and install. Nevertheless, Chapter 14 will deal with *their* limitations.

Chapter 7

What's going on? Recording levels and metering

7.1 The transition from analogue to digital

Practices and traditions in the recording industry have usually been historically dictated by good engineering principles and available technology.
Sometimes, however, even when a new and radical technology has become
available, it has been difficult to shift many of the old ways. Unfortunately, in
many instances, as with religions, time often causes true understanding to
give way to blind faith, and 'Because this is the way it is done', becomes the
stock answer, rather than 'This is the reason *why* it is done'. The change from
analogue to digital recording systems seems to be such a case, where analogue metering practices have been applied 'in their entirety' to many digital
recording systems for which they may not be appropriate. Lack of understanding of the needs of analogue metering has tended to lead to the view that
'a meter is a meter' which has allowed the inappropriate use to go unchecked.
In fact, there is a wide range of music signal metering for many different purposes, so perhaps we should look at a few.

7.1.1 The traditional approach

Traditionally, tape recorders in professional use have been treated as unity
gain devices, putting out as close a replica as possible of what was put in, both
in absolute level and frequency balance. Traditionally also, machines have
had input and output controls, facilitating the variation of actual magnetic
levels recorded on tape, either for reasons of the dynamics of the music, or to
gain optimum performance when changing from different makes or types of
tape. Let us say that a manufacturer brings out an improved, higher level tape,
accepting an extra 4 dB, yet with all other parameters remaining more or less
constant. Conventional practice would require that the reproduce level controls be backed down by 4 dB, thus reducing tape noise and reproduced signal
by a similar amount. The record level control would then be increased by 4
dB to restore the unity gain through the machine. The result would be that the
recorded signal would be 4 dB higher on the tape, and thus the reproduced
signal would see a 4 dB improvement in the signal to noise ratio, but the
machine would still remain a unity gain, '0 in/0 out' device.

With such set-ups, a mixing console putting out a nominal 0 VU (1.23 V)
would show 0 VU on its output meters, and it would also show 0 VU on the

tape machine meters. When the reproduction from the machine was selected on the console monitors, 0 VU would be seen on the outputs meters of the tape machine, and this level would return to the console. This would be irrespective of the actual magnetic recording level. In the two cases described in the previous paragraph, there would be no difference between the signal levels entering or leaving the machine, nor in the metering levels. Only by reference to a standard reproduce alignment tape (test tape) could the actual magnetically recorded levels be determined. When aligned in this way, it is easy to 'A/B' between the mixing bus outputs and the reproduced output from the machine, albeit delayed by some milliseconds due to the distance between the record and the reproduce heads, and a very accurate judgement can be made of any loss of quality or other artefacts produced by the tape recorder or the tape. If for example, on A/B-ing, it was found that a piano recording was saturating, then either the level from the mixing desk could be reduced, or the level on tape could be reduced by the reverse of the process described above. The second option would still leave the tape machine as a unity gain device, but would also leave the mixing console operating at *its* optimum level for the best noise/distortion compromise.

Given the 'soft' limiting characteristics of analogue tape, typically 20 dB of headroom would be provided in a good mixing console to allow innocuous peaks to pass cleanly to the tape recorder, where, dependent upon the nature of the signal, they would probably be rounded off without too much audible effect. Were the mixing console to have the same overload point as the nominal tape overload level, then the harder, more aggressive electronic clipping of the console outputs would be most undesirable. As neither PPMs (peak programme meters) *nor* VU meters ever truly represented any universal maximum which could be allowed to pass to tape, it was always a case of the experience of the engineer interpreting the meter readings and allowing a level to pass to tape based on his or her experience of the characteristic sounds of different instruments. (Incidentally, VU stands for Volume Units, an old telecommunications quantity relating more to perceived loudness than to absolute level.) There were no absolute rules for the interpretation, either, because each type of machine, each mixing console, and each brand of tape had their own individual overload characteristics. What is more, as headroom tended to be expensive to achieve, semi-professional, less expensive equipment tended to be much less tolerant of overloads, therefore required much more care in the judging of optimum recording levels. Ironically, such equipment normally had cheaper, less accurate metering.

When viable digital tape recorders made their debut in the 1970s, they were made, externally at least, to be as operationally similar as possible to their analogue counterparts, partly to ease their acceptance into an industry with complex but well-established ways. The primary differences for a recording engineer were that the digital machines had a much greater margin between noise and distortion, hence a greater usable dynamic range, and also that the distortion, once it did begin, was totally unacceptable; there was no soft clipping like the analogue machines. The increased dynamic range began to highlight noise problems elsewhere in the recording chain, where hitherto unnoticed problems which had been masked by the vagaries of analogue now had to be addressed. Apart from that, the transition from analogue to digital recording was a relatively painless process.

7.1.2 The digital influence

By the mid 1980s, digital technology had begun to penetrate deeply into the domestic audio market, and unlike the domestic/professional differentiations of analogue, it was soon realised that a bit was a bit in digital terms, and no bit could know whether it was recorded on a professional or a domestic format. In other words, once in the digital domain, free of A to D and D to A convertors, no intrinsic difference existed between a signal recorded on a domestic, semi-professional *or* a professional machine. Thus the R-DAT format, despite being heavily stressed by its inventors, Sony, as being for domestic only use, was hi-jacked by the semi-professional *and* professional recording industries as a very inexpensive means of digital mastering. This was where a great nonsense began to enter the operations.

Clearly, in digital terms, the absolute maximum peak level before clipping is the optimum level to record, as this will preserve the greatest amount of detail in the low-level signals. Nothing is to be gained on *any* musical signal by using a lower than maximum level on any digital tape, disc, or whatever other storage medium which may be used. But, and it is a big but, most DAT machines do not have reproduce output level controls, and most of the modular digital 8-track machines have neither input *nor* output controls.

Let us take first the example of a Panasonic SV3700 'Professional' DAT recorder. (I put 'professional' in inverted commas because, although the word appears on the machine, to the best of my knowledge the whole format is still not intended to *be* professional. However, this machine does make a big effort to justify its label.) If we connect this machine to the stereo mixing bus output of a conventional professional mixing console, and put out a maximum peak signal at 0 VU, (1.23 V/+4dBv) then good digital practice would dictate that we set the record input control of the DAT machine to one light below clip on its record level indicators. The operation manuals of most DAT machines are remarkably vague on the subject of recording levels, but *if* we do this, recording on the verge of distortion, we will be optimising our recording level. Remember, in digital terms, 'on the verge of distortion' is still as pure as the driven snow, but what about in the analogue circuitry to which it is all connected? If we put a tone out of the mixing console at 0 VU, then adjust the record input control of the DAT to give 0 VU *out* of the machine (which is the typical set-up in terms of analogue machines), then we find that a signal level of around 15 dB below clipping gives us a unity gain machine. In the absence of reproduce controls, this would suggest to me that the manufacturer intended that –15 dB would be the intended recording level. In reality, the reason for this is, no doubt, that the machines are still trying to follow an old analogue recording concept, even now that digital is fully established in its own right. They are still trying to be 'user friendly' to people whose only other recording experience is from analogue systems, but this is hardly the sort of practice to get the best out of digital systems. We are throwing away 15 dB of dynamic range to allow the machine to work with a mixing desk which only has VU meters.

If a mix is being made on an analogue mixing console with standard VU meters, then a snare drum or a grand piano, visibly peaking at 0 VU on the meters, could actually have peaks around 10 or 12 dB higher. As mentioned before, they would soft clip on analogue tape, and it would be down to the engineer to decide whether that amount of tape compression was acceptable, or not. This is consistent with the 15 dB of apparent headroom available on the standard set-up of the DAT machines. It also highlights the workable correlation between VU meters and analogue tape, at least in the hands of experienced recording engineers, and also explains why such a crude device as a VU meter has now seen over 50 years of professional use. The truth is, though, that VU meters have very little relevance to digital recording. The question therefore seems to be 'should the DAT machine manufacturers be compromising the capabilities of their machines by adhering to the operating practices of a metering standard which is not suitable for digital recording?' If we disregard this, however, and still go for our 'optimum' one light below peak, the 0 VU tone *into* the machine will give 15 dB *over* 0 VU *out* of the machine, or +19 dBv. Apart from the fact that one must question if the output amplifiers of the machine are happy with such high levels, that is, are they at their sonic optimum, almost certainly the monitor inputs of many mid-priced mixing consoles will not be too happy with +19 dBv input levels. What is more, any meters on the machine returns of the console will be almost permanently pinned to the end stops, and inadvertently pressing the buttons to select playback, to hear the tape after performing a loud mix, may risk ruining the loudspeakers when the mix comes back 15 dB higher in level than it left the stereo group outputs. As few digital machines can reproduce whilst recording, A/B, line/tape monitoring is not really possible, but even if it was, it would be rendered impracticable by the 15 dB mis-match from the levels entering and leaving the machines.

7.1.3 Level interface of MDMs

A tone at 0 VU (+4 dBv) into the balanced inputs of machines, for example, will give a signal level indicating 18 dB below peak on the ADAT meters. In turn, this will produce 0 VU at the balanced output terminals, which follows the general operating convention of normal analogue machines. Again, if the philosophy of utilising maximum digital recorded level is to be adhered to, then a signal level of +22 dBv (+4+18) would be needed from the output of the mixing console. Almost no mixing consoles have meters which will read these levels, so one must then rely solely on the ADAT meters. The first time that I used the ADATs, I was a little worried when the odd peak of an unrepeatable live performance briefly flicked the red lights on. I was both relieved *and* surprised on playback to find that despite the peak indicators still indicating an overload, no audible distortion was evident, so these peaks must have been something like +21 dB over 0 VU, or +25 dBv at the output. Obviously, on reproduction, as an ADAT is a unity gain device, these will be the peak levels entering the line inputs of the mixing console. It is true that with many machines, the unbalanced −10 dBv outputs could be used on mixdowns, but nonetheless, even *those* outputs may not be very happy reproducing +11 dBv or more: 21 dB above their nominal operating levels.

7.2 Differing meter philosophies

We seem to be entering an area where optimisation of *digital* levels is done at the expense of marginalising the performance of the analogue audio circuitry, which is surely a nonsense if we are seeking to achieve improved *overall* optimisation of the record/reproduce chain. Interface in general seems to be becoming more and more arbitrary. I recall a strange occurrence when a colleague had to make a digital transfer from a Panasonic SV3700 to a Fostex D20. The Panasonic was reading peaks of one light below overload, whereas the Fostex was clearly indicating overload. Once again though, on playback of the copy, no distortion was noticed. For practical reasons, digital recorders drive their meters from the digital signal. This is necessary to allow a direct digital input to show on the meters. The chosen internal headroom may vary from one manufacturer to another, and therefore, so do the meter readings when copying. This occurs either via analogue interfaces or when transferring digitally from one machine to another. I have also heard comments that many engineers were only loading maximum on to DAT, in the belief that many CD plants are not capable of boosting a digital signal level, so a low level DAT master would render a low level CD. They were thus using the tape optimally but for an entirely different reason. There is great uncertainty around the industry, created by some very unclear instructions from equipment manufacturers about how they intend their machines to be used, and there is almost no information as to why they adhere to their respective philosophies. This situation is not much help to anybody.

I also wonder just how 'fast' some metering circuits are. Clearly some of the level indicators on DAT machines indicate overload at radically different levels; perhaps 6 dB or more between indicated overload and audible overload when going from machine to machine. The fact is that for marketing purposes, the manufacturers are attempting to establish a digital/analogue machine interchangeability, no doubt to ease the up-grade process, but the two are very different systems, and digital recorders often cannot function on analogue orientated recording chains without a serious loss of their potential capabilities. (Well, not without record and reproduce level controls on the recorders, that is.)

7.2.1 Dynamic ranges

Let us consider the comparisons in more detail. Figure 7.1 shows the dynamic ranges, side by side, of a high-quality mixing console, together with representations of the dynamic ranges of an analogue tape recorder and a digital tape recorder. The mixing console and the analogue tape recorder have an optimum, 'zero' level, clearly marked. It will be noted that the console has a clip level and a noise floor which are not only separated by a much greater dynamic range than between the noise floor and reference level of the analogue tape recorder (usually the level where the onset of tape compression begins to render harmonic distortion greater than about 1 per cent), but also between the noise floor and *saturation* level of the analogue tape recorder (where gross distortion begins). This means that the console reference level can be positioned somewhat arbitrarily, as long as it yields a console clip level which is above the saturation level of the analogue tape recorder, and a noise

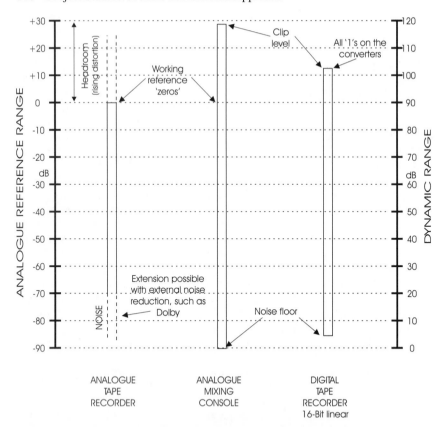

Figure 7.1 Dynamic range comparisons. The dynamic ranges of the devices are somewhat arbitrarily positioned alongside one another, but in such a way that would be typical in many professional recording systems. In each case, the upper limit is set by the onset of unacceptable distortion levels, and the lower limits by the noise floors

floor which is below that of the noise floor of the tape recorder. In practice, this is not too difficult to achieve, because it is much easier to make console electronics with 120 dB of dynamic range than to get more than 90 dB from an analogue tape recorder, even with the use of external noise reduction devices.

On some mid-priced consoles, the upper half of the range between the nominal '0' output level and the clip level may begin to become of lesser quality than the performance up to that level, but as analogue tape is beginning to saturate, its distortion products will usually be higher than those of a slightly degrading console output, so this will tend to mask any deficiencies in the console outputs at levels over, say +10 or +14 dB relative to the nominal '0'. Top of the market consoles, designed to operate at a nominal +4 dBv (0 VU) output level, would typically deliver output levels of +24 dBv, or even higher, without any signs of sonic degradation, but that may well not be the case for the many consoles designed for the middle market, or project studio use. Nevertheless, with analogue tape recorders, they function well enough for this limitation not to be a problem, thanks to the aforementioned masking effect.

To the right of the console representation, in Figure 7.1, we have the dynamic range representation of a typical digital tape recorder. It is limited at its lower level by the noise floor of the D to A (digital to analogue) converters, and at the highest level by the sudden onset of gross overload distortion. At no point in between those upper and lower limits is there any significant difference in the distortion produced by the digital recorder, although at very low levels, quantising artefacts may become apparent. Figure 7.2 shows the typical distortion to level relationship between digital tape recorders, analogue tape recorders, and mixing consoles, although here we see two different mixing consoles, one top of the line, and another of more modest performance. This demonstrates the excellent correspondence between the analogue recorder and both of the mixing consoles when aligned and matched conventionally. This is not really surprising, as they have developed side by side for the past 50 years.

As we have no preferred level, from the point of view of distortion, for recording a tone on to a digital tape recorder, it follows that our main consideration, in terms of recording level, will be dominated by considerations of noise, which will determine the limit of the dynamic range of any recorded signal. Hence, by recording our music signals with peaks just below the converter clip level, we will maximise the dynamic range, and hence the resolution of important low level information. The further we can keep the low level signals above the noise floor, the greater will be the sense of openness, cleanliness and dynamism of the music. In fact, there are schools of thought which define the maximum recorded level to be that which just causes imperceptible distortion on peak signals. The perception of peak clipping is a very complicated subject, but if any given levels of clipping are imperceptible, then whatever those levels may be, they will not be detrimental to the music. This could have been happening in the previous reference to recording to a point where the overload lights had illuminated, but no distortion was heard on playback.

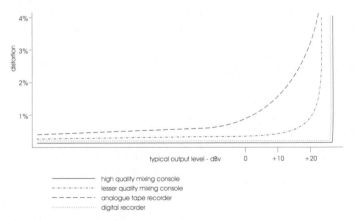

Figure 7.2 Distortion level comparisons. It can be seen from this plot that there are different characteristic shapes of 'knee' at the onset of distortion. As the 0 to + 20 dB range can still be usable with analogue tape recorders, such as when recording fuzz guitars or some percussion, the use of this range of 'headroom' in an analogue console is still beneficial. With digital recording, however, the distortion onset characteristics of the system are almost identical to those of a good mixing console, so the headroom concept has little relevance if suitable peak metering is used.

7.3 The roots of distortion

Perhaps we should now look at what the mechanisms for distortion are in each of the three systems under discussion. A mixing console will overload relatively suddenly when the peak magnitude of the signal voltage, at any part of the system, reaches the value of the supply rail(s) from the power supply. A system running on ±5 volts DC, cannot produce an output of more than 10 volts, peak to peak, of music signal. In practice, for numerous reasons, the signal may begin to degrade before the absolute limits are reached. An analogue tape recorder records both the varying signal current and a fixed, high frequency bias current, which helps to offset the hysteresis problems in the tape (which does not magnetise and demagnetise symmetrically). Tape will gradually begin to compress when the range of relatively linear recording provided by the bias signal has reached its limits. If the music signal exceeds the bias signal, the effect of the bias is lost, so the tape saturates rapidly, and non-linearly. However, gradually increasing tape distortion is not as harsh as with the hard clipping of most electronic systems, because much of the distortion consists of even order harmonics which can sound quite musical. The effect is more like that of limiting, which gradually becomes harder as the level rises.

In the case of digital recorders, it is not the tape which overloads, at all, but the analogue to digital converters. All the tape is ever doing in a digital recorder is recording coded numbers. It records the same overall number of digits, irrespective of the musical signal level. Once the converter reaches all '1s', though, there can be no more, so aggressive sounding hard clipping results. Digital overload therefore has nothing to do with signal levels on tape, but only with running out of numbers in the A to D converters. The above three mechanisms of overload are thus vastly different, so it is little wonder that optimally matching the systems can be problematical.

Figure 7.3 shows what happens when we compare the typical interfacing of a middle priced mixing console to a digital tape recorder. In Figure 7.3(a) we see that with a mixing console using VU meters, and a signal having peaks of 6 dB above the 0 VU level, then if the units in question, say, a console and an ADAT, are connected with the +4 dBv output of the console connecting to the +4 dBv inputs of the ADAT, the peak levels will still only reach 12 dB below clip on the ADAT. This leaves two whole bits of unused dynamic range. (ADATs clip at +18VU. Six dBs equals one bit of resolution.)

In Figure 7.3(b) we see what happens when we decide to increase the console outputs so that the ADAT now receives a peak level at its clip point. The console is also very near to its clip point, and the distortion, at this level, may be rising. With the digital recorder, unlike the analogue recorder, there is no gradual increase in distortion as the level rises, so there is no masking of the increasing distortion from the console. If it exists, it *will* be heard. Furthermore, with this type of connection, driving the console output to +22 dBv will be well off the scale of the meters, so they will perform no useful function. Likewise, when the output of the machine returns to the desk, the meters will also show full-scale, and the input circuitry, on receiving +22 dBv, may, depending upon circuit design, suffer similar stress to the output stages.

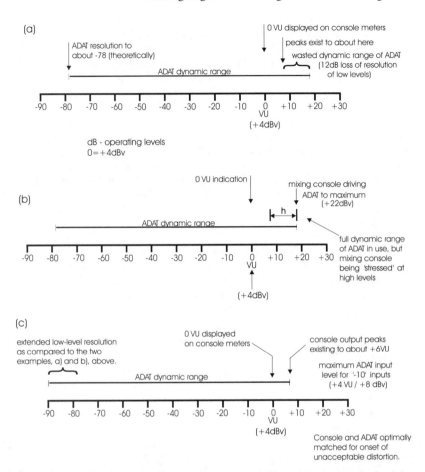

Figure 7.3 (a–c) Interfacing considerations. (a) ADAT with a typical mid-priced mixing console (low level signals compromised). This assumes that the +4 dBv inputs and outputs are being used on both the console and the recorder. With peaks of +6 over 0 VU, the console would be likely to be operating about 8–10 dB below the onset of gross distortion, but the sonic neutrality may begin to deteriorate if much higher levels than +6 VU are used. The example here is therefore optimal for the mixing console, but wastes about 12 dB of low level resolution in the recording system. (b) As (a), but driving the console to its limits (high level signals compromised). This operating regime would utilise the ADAT dynamic range to the full, but signal levels in region 'h' would be off the scale on the meters of most mixing consoles, rendering them useless, and on many typical mid-priced mixing consoles would be in the region of questionable sonic neutrality, as electronic circuitry was 'stressed'. (c) Use of '+4' console inputs and outputs with '–10' ADAT inputs and outputs to achieve maximum dynamic range of recording system and viable metering.

7.3.1 Circumventing console distortion

There are two simple solutions to this problem. The first is to use the –10 dBv inputs and outputs of the ADAT with the +4 dBv inputs and outputs of the mixing console. By this means, as shown in Figure 7.3(c) +4 dBv from the mixing console, will produce a signal 14 dB above the nominal input level of the recorder, but, as the nominal input level reads +18 dB below overload, on tones, this would still yield 4 dB below clip on the ADAT. This is just about perfect, because it gives a realistic 4–6 dB margin for musical peaks, above the 0 VU of the mixing console meters. Greater care should obviously be taken if the console only has conventional VU meters, in which case it would be better to rely on the machine meters. The only drawback to doing this is that it means using the low level, unbalanced machine inputs and outputs. If the machines are some distance away, such as in a machine room, then this may render the system more prone to noise pick-up in the lines, but if there is no apparent noise problem, or the machines are sited close to the console, there should be no problem to using the lower level, unbalanced inputs. In fact, on some machines, it is arguable as to which inputs and outputs (the low level, unbalanced, or high level balanced) are the most sonically neutral. Most modular 8-track machines tend to be built to a price, remember, and high level balanced inputs of great sonic purity are not usually cheap items to make. It is therefore not surprising that if line noise is not a problem, the low level, unbalanced terminations may actually be sonically preferable to any financially compromised, high level, balanced interfaces. This question may ultimately be a function of cable length, though, and can only be judged by careful listening in each individual case.

7.4 The optimum solution

Of course, this whole situation could be avoided by a second option. Simply by putting input and output level controls on the digital recorders they could be made optimally compatible with their analogue counterparts, *and* the mixing consoles. In fact, this is such a fundamental requirement that without it, there is no absolutely optimum way to interface these machines. So, once again, their claims to any professional status is put in doubt. Of course, one of the roots of this problem is still the metering dilemma, so perhaps we should take a look at the meters in current use, and how they came into existence.

7.5 Metering

The standard VU meter dates back to 1939, and VU stands for Volume Units. In fact, the device is really called a Volume Indicator Instrument, but rarely gets called that, these days. It was designed to monitor the loudness of signals, on telecommunication lines. VU meters are not simple meters, but are specially designed to respond to the average level of programme. The ballistics are damped such that the overshoot levels on peaks are reasonably well related to what is heard. VU meters do not relate well to rapid peak levels, so they are not much use in situations where it is important to monitor absolute peak levels, such as with radio transmitters or vinyl disc cutting equipment. In both those situations, the passing of unseen peaks can lead to some very expensive equipment repair bills. Nonetheless, the VU relates reasonably well to analogue tape saturation characteristics, and recording engineers soon learn to correlate the meter reading with the type of sound being recorded, and use levels which correspond with what their experience tells them.

For example, with an optimally lined-up tape machine, an instrument with fast, clean peaks, such as a well-struck piano, would perhaps be happy peaking around –4 on a VU meter. A fuzz guitar, on the other hand, with virtually zero difference between peak and mean signal levels, and which is already 'distorted', could well be recorded around the end-stop on the VU. The fact is, though, that the metered levels need experienced interpretation. Cheap VU meters on inexpensive, or even modestly priced mixing consoles are unlikely to have true VU ballistics, so the interpretation of their readings may take some degree of trial and error. High quality VU meters are expensive, and hence only tend to be found on expensive equipment. Here, though, by 'expensive' I do not simply mean 'equipment that costs a lot of money'. To most people, $50,000 is a lot of money, and for that amount of money, a 16-channel mixing console with few facilities *would* be expensive. On the other hand, at the same price, a 96-channel console with full equalisation and many extras, would *not* be expensive. Only the former would be likely to have true VU meters.

VU meters, however, relate only very poorly to digital signal levels, perhaps to the extent that Alesis felt it necessary to leave an 18 dB safety margin on the ADAT system. That may have been a marketing solution, but matching the machine to typical mid-priced mixing console concepts is clearly not a technical one. One may be tempted to think that peak programme meters (PPMs) would be more relevant, but the question would be, 'Which PPMs?' In Europe, alone, there are at least four types of PPM in use. The European Broadcasting Union, the German DIN Standards Institute, the Nordic Broadcasting Authorities, and the UK (BBC) all have their own ideas of what a PPM should be. What is more, in the USA, even some individual broadcasting companies have their own, in-house, PPM standards. Over all these different PPM standards, there is a very wide range of scale differences, as shown by the examples in Figure 7.4.

In response to this great confusion, and just to add some extra spice, many manufacturers of digital audio equipment have introduced their own metering standards. There is currently no general consensus. The problem is compounded by there being no simple correlation between audio signal levels and digital signal levels. The digital signal 'levels' are largely concerned with how many bits are left, which relates to the *content* of the audio signal, as much as to its absolute level. Several organisations have decided that A to D and D to A converters should be calibrated such that 0 dBFS (0 dB, Full Scale – i.e., all '1s' in digital code) should refer to +18 dBv, but this is, again, to ensure compatibility with analogue systems. This is an obvious compromise, because it does not take into account the full potential of digital recorders. It is another fudge.

The only real solution to the metering problem seems to be to use a remote metering unit, at the mixing console, which is manufactured specifically for the digital recorders which are being used; unless, that is, the machine meters can be seen easily from the console. Once again, though, as with all meters, a little bit of trial and error is involved in order to gain experience as to just how 'hot' things can go on different types of signals. The illumination of the clip lights on violin recordings is not recommended, but I have seen many a

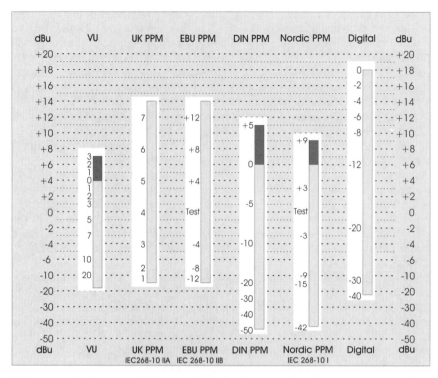

Figure 7.4 Some commonly used, widely differing meter scales. The low-level end-scale is non-linear so spacing may not appear as shown. The numbers shown on these scales indicate the general intervals but not the precise indicators defined by the relevant standards (after Peter M. Harrison)

percussion instrument cause the red lights to flicker on digital recorders with total impunity, because the ear cannot recognise distortion unless it is sustained for a sufficiently long period of time.

Metering is a guide, and only a guide, to what is occurring in the signal chain. As yet, there is no substitute for experienced interpretation of their readings if optimised recording levels are to be routinely achieved, and this fact must be borne in mind no matter what type of recording system is being used.

7.6 Digital to digital metering

So much of what we have been considering in terms of metering problems have been how the differing systems relate to each other. One implication of this should be that by using similar metering systems on either wholly analogue or wholly digital recording systems, all should be straightforward. In other words, if we monitor a signal level on a digital mixing console, then digitally connect it to a digital recording device, be it hard disc, CD, tape, or whatever else, we should have no discrepancies. Well, it is true to say that, in general, things do correlate quite well, however, from time to time we can observe some misrepresentations.

Obviously, with digital to digital signal interconnections there is no need to take into account any of the headroom aspects which need to be allowed for in analogue recording systems interfacing. There is no risk of gradual signal degradation as levels rise, and there are no saturation problems to consider. On the face of it, therefore, as a bit is a bit in any digital system, there should be absolutely no problem in making meters read to any given standard.

It was mentioned in Section 7.2, however, that different levels had been seen on the meters when sending one signal to two DAT machines from different manufacturers. Remember that digital 'level' meters are not actually measuring sound level, but sample rates and levels in the digital domain. At low frequencies, say below 1 kHz, this tends to relate reasonably well with the actual musical signal level, but at high frequencies, and especially with sine waves, all manner of nonsenses can result. In fairness, it must be said that not too much music uses full-level 12 kHz sine waves, but such signals can provoke some very strange readings with some digital metering systems. Odd situations can occur whereby some signals, no matter how highly they are recorded, may never be able to show an overload. This is the result of the different manufacturers having different ideas of what best approximates to most musical signals. Their different opinions of exactly where the subjective perception of overloads begin lead them to different metering solutions.

Different opinions also exist as to how best to compromise to human eye response times, and to what degree to maintain metering compatibility within the range of equipment of any one manufacturer. For example, one very well-known manufacturer developed a metering system for a one of its mixing consoles which gave excellent correlation between actual digital 'levels', the music and the human perception of such metering. However, the meters gave noticeably different reading to the meters on the company's tape recorders. Even though they were more accurate in almost every way, the new meters were scrapped, and the consoles were fitted with the inferior meters which corresponded with those on the tape recorders. This was purely a marketing decision, to avoid any embarrassing questions from users. As these were fully professional and very expensive pieces of equipment, I think that the decision was a disgrace, but such is the current climate.

With this situation existing, even at the fully professional level, one can hardly be surprised that the uncertainties at the semi-professional level are even greater. Ironically, though, it is the equipment in the hands of the less experienced operators which really needs the *most* accurate metering, because the wherewithal to make the best interpretations of the readings is less common.

So, even in all-digital chains, there is no guarantee that simply watching the meters on one device will necessarily suffice. Different devices may show overload points up to 4 dB apart, and these may also be signal dependent. With experience, and on some regularly used chains of equipment, users may become familiar with the metering, and learn which meter in the chain is most sensitive, *or* most relative to the onset of perceivable distortions. Once such experience is gained, it may then only be necessary to pay attention to one set of meters in a chain.

Digital to digital metering is *not*, by any means, as absolute as it perhaps ought to be in an ideal world, … or a *truly* professional one, for that matter.

Chapter 8

Mastering

The tendency amongst project studios is for them to use a wide range of different monitor systems, and for them to exist in a wide variety of inadequately controlled rooms. More often than not, the control rooms are neither expertly designed nor built to rigorous acoustic principles, so the variability in monitoring conditions, from room to room, can be considerable. To a somewhat lesser degree, variability exists from room to room even amongst the more elite studios, so there is a need, somewhere in the production process, to pass the recordings through the hands of a set of people who can be objective about what they hear. These people are the mastering engineers, who can help not only to act as a last line of quality control before the recordings go to the manufacturing plants, but can also use their experience to make recordings from a multitude of sources converge towards some sort of standards of tonal balance. This is obviously desirable if manufacturers of domestic loudspeakers are to have any decent production responses to aim for.

Much of what is included in this chapter grew from an attempt to find if there was any general consensus amongst specialised mastering engineers on the subject of preferred monitoring systems. In fact, there was very little in terms of which brand or model, and yet the better mastering houses manage to turn out very consistent products, despite their diversity of equipment. This strongly reinforces George Massenburg's comments which were quoted in Chapter 5, to the effect that monitors either work for an individual, or they do not. I suppose that the same could be said about tennis racquets. Champion tennis players do not all use 'the best' racquet, they each use the ones which help them get the best results. In the end, though, it is all down to the skill and the experience of the players, and the choice of racquet may be down to finding the one which most suits the style of play of each individual contestant. However, at championship level, it goes without saying that all the racquets must be of high quality. This analogy has very close parallels with mastering engineers and their choice of monitoring systems, although as with the tennis players, it is the skill of the person using the equipment that ultimately achieves the results.

Though there may seem to be no reason why a recording cannot be mastered in the studio where it was recorded, provided that they possess good monitoring conditions and the necessary equipment, it should become apparent whilst reading this chapter that many of even the most experienced engineers and producers prefer to pass their recordings through the hands of trusted and experienced mastering engineers, even if, subsequently, nothing is changed. There are psychological, as well as technical and artistic reasons for this, so, let us begin with a look at the way in which the mastering profession has developed, and hear some words of wisdom from some of the well-established providers and users of the services.

8.1 The evolution of a profession

Mastering is the last link in the chain of assessment of recorded music before the end result is unleashed on the music buying public. Historically, this final assessment was performed by disc-cutting engineers, often in conjunction with the recording engineers and/or producers who had performed the mixing in the studios. What could be put on to tape in the studio was often much less restricted than what could be put on to disc, with what could be taken *off* disc being the *most* limiting factor. It was thus of no use whatsoever to mix a recording without giving due consideration to whether or not it could be played back on an average 'record player' when the final product was sold in the shops. A recording which would cause 'the needle to jump' could cause a great deal of expense to the record company due to returned discs, as well as affecting the reputation of the artiste(s). The disc-cutting engineers were the people with the specialist knowledge of how far a producer could push things in terms of overall level, stereo effects, low and high frequency boosts, and a number of other limiting factors, before problems were likely to occur. Recording engineers, producers and disc-cutting engineers often used to form very close working relationships, and the partnerships usually formed around teams who could get the best mutually agreeable compromises.

With the advent of tape cassettes a different set of constraints was added, due to the limitations of the high-speed duplication process, but in general, if something could be reproduced from a disc on an average quality record player, then it would be very likely to go quite happily on to a cassette. More recently, though, compact disc has become the predominant medium for recorded music, with an inherent performance capability equal to almost whatever the recording studios themselves can produce, and this has now thrown the burden of response limitation on to the domestic loudspeakers and amplifiers.

8.1.1 Vinyl limitations

In the case of vinyl discs, high levels of low frequencies and out of phase signals have always been a problem. If the low frequencies, such as bass drums and bass guitars, are positioned towards the centre of a mix, they will produce a lateral movement of the cutting head. The greater the amount of low frequencies, the further apart must be the adjacent turns of the groove, as more low frequencies cause the groove to wiggle to a greater degree, and adjacent turns could risk bumping into each other if not adequately spaced apart. The 'lines per inch' spacing ultimately decides the total amount of time which a disc can support, so high levels of low frequencies can determine the overall playing time of the disc. If the playing time required from one side of a 12 inch LP is above 20 minutes or so, lower than normal levels of signal will need to be recorded in order to allow more lines per inch to be cut. It is critical, however, to keep as high an average level as possible, as surface noise is always around to blight low level signals.

As discs rotate at a constant 33⅓ revolutions per minute, the actual speed at which a stylus travels along the groove will gradually reduce as it moves towards the centre of the disc. One, of course, has to stop the music somewhere short of the centre label, or perhaps more importantly, short of where many automatic lift-off systems operate. Sometimes, however, depending on the precise musical signal at the end of a recording, it is necessary to stop well short of the label or lift-off point if the specific requirement of the musical signal cannot be supported by the gradually slowing, relative groove/stylus speed.

It is often unfortunate that the vinyl disc had not been standardised to run in the opposite direction, beginning in the centre and running to the outside, because the requirement of much music, with quiet beginnings and loud crashes at the end, would have been better served. In fact, centre-start discs *were* used for broadcast transcription until after the Second World War, but the other system prevailed. After the oil crisis of 1973, virgin vinyl became very expensive, and much re-cycled vinyl began to be used. Around the same time, hand loaded presses began to give way to the faster, injection moulding process. Instead of virgin vinyl chips being evenly spread over the stampers, and the steam heated presses producing a homogeneous disc, reconstituted liquid vinyl was pumped into the centre of the already closed presses. Any dross or general impurities were forced ahead of the spreading vinyl, rendering a more polluted plastic around the region of the run-in groove. Consequently, with a conventionally rotating disc, the outer grooves tend to be better for the loud crashes and bangs, where the faster tracking speed can more effectively support the high level transients, and the higher recorded level can swamp the surface noise from the poorer vinyl. The quieter beginnings would enjoy a better signal to noise ratio if recorded in the more pure, more homogeneous vinyl near to the centre of the disc, and where the typically less demanding musical signal could be easily accommodated by the slower tracking speed. Anyhow, discs run from outside to inside, so the cutting engineers have had to learn the limits very well indeed, and cut the disc with a great deal of care and compromise.

High levels of out of phase signals, especially those caused by putting instruments with high levels of low frequency content to one side of the mix, create large vertical movements of the cutter head. The maximum cutting depth must be set so that the cutting stylus will not destroy itself by hitting the aluminium core of the lacquer, the upper limit being where the groove ceases to exist as the cutter loses contact with the acetate surface. As all playback cartridges have a stylus of a finite mass, they will consequently also possess momentum when accelerated, so one cannot allow the stylus to be launched into the air with too abrupt a vertical movement of the groove or it will continue to follow its own ballistic trajectory, and will not be under the control of the groove. Only some of the finest cartridges have usually been able to track very high level transients. Almost always therefore, because of these limitations, compromises to the domestic market have usually meant that the audiophiles have had to suffer some loss of potential overall quality from a commercially viable disc.

In the above paragraphs, I have only just touched on a few of the problems which disc-cutting engineers have had to face, so it can be seen that a very great number of factors have to be balanced and well judged for a best overall compromise disc to reach the shops. In general, disc-cutting engineers have been a rather level-headed group of people: they have had to be! They have been so constrained by physical processes that there has been no room for the creative excesses which are more readily found in recording engineers and producers. The cutting engineers jobs were always at risk if shops and pressing factories returned too much of their work, and they never had any of the 'artistic' get-outs of the recording staff. What is more, cutting heads have always been as delicate as they were expensive, and, in the years past, a blown head could cost the equivalent of three or four months' wages of a cutting engineer. Inevitably this led to a situation where only trained professionals were allowed to operate the lathes, so automatically this ensured that an experienced pair of ears was present to assess the last stages of the production process. Basically with the CD, no such restrictions apply, so the tape master need not make compromises to the final storage medium. In turn though, the removal of the vinyl-imposed restrictions has led to a bit of a free-for-all. It suggested that the studio tapes could go directly to the CD factories, which in turn has led to much general unawareness of many potential problems. Remember, when vinyl was the limiting factor, it served to protect the domestic system weaknesses, but the removal of the restrictions has allowed many bad practices to now pass to the end product. The discipline which vinyl imposed served as a useful teacher to previous generations of recording personnel.

8.1.2 Life after vinyl

When vinyl ruled, the channel separation was never more than 30 dBs or thereabouts, so a guitar panned hard left on the master tape would 'bleed' into the right channel, around 30 dB down. With analogue magnetic tape the channel separation is in the order of 60 to 80 dBs, so, what is panned left is unlikely to appear above the noise on the right. With digital recording media, the separation is effectively total. On loudspeakers the effect of this change is not dramatic, but on headphones, hard-panned mono signals can sound very thin, just as the whole sound loses body if one listens to only one side of a headset. The bleed across the vinyl pick-up supplied enough cross-channel signal to keep a sound such as a hard-panned guitar sounding full. Consequently, it seems unwise, nowadays, to hard-pan a mono signal source, as the headphone balance may then become unacceptable. A small amount of cross-feed to the opposite channel will not affect the perception via loudspeakers, but it will ensure that left and right signals remain full-bodied when heard via headphones. Such things are caught by sharp mastering engineers, but in the studio they are often not thought of these days.

Concurrently with the demise of vinyl came the almost interminable supply of signal processing effects. Many of these cause or rely on phase effects and multiple delays, which can also produce significant out-of-phase content in the overall mix when used for stereo effect. I believe that much of this is at the root of the problem as to why so many modern mixes sound unnaturally hard. In the days of vinyl, these things would have to be corrected before the cutting, but now, such constraints rarely apply. To avoid a state of total chaos, at least *somebody* in the recording process should be keeping an eye on these things, and these days, that job usually lands at the door of the mastering engineers.

8.1.3 The advent of the mastering engineers

Many of the current mastering engineers were formerly, or still are when needed to be, disc-cutting engineers. They have been trained in the disciplines required to achieve the optimum performances from whatever medium for which they may be mastering, or the best compromise for multi-format releases. During the days of vinyl, to pass something past their ears was almost mandatory, as few recording engineers of the last 25 years have been able to cut discs themselves. The cutting/mastering engineers' more detached and perhaps more objective assessment of the master tapes was usually a valid extra check that all was well. They would listen more in the way that the record buying public would listen, without any knowledge of the trials and tribulations which had existed during the recording process, yet at the same time they could offer specialist advice.

I well remember sometimes making five or six mixes of a recording, each time making finer and finer adjustments until an acceptable ultimate mix was produced. The tape box would be clearly labelled as to whatever made each of the earlier mixes unusable; 'The piano is 1 dB too quiet', 'The guitar fades in too slowly' and so forth. Six months later, I could listen to all of the mixes without preference, yet in the intensity of the mixing situation, I was convinced that the first mixes were useless. The more matter-of-fact judgement, which I could only make after sufficient time had passed to allow me to forget the nit-picking detail is something which a well-trusted mastering engineer can supply immediately.

There is, of course, currently no absolute need to use the services of a mastering engineer, as many recording studios now have all the necessary editing equipment to produce a finished digital (or analogue) master. However, there is a totally ludicrous disparity of the monitoring conditions from studio to studio, and an even greater proliferation of more or less self-taught recording personnel. It is therefore not surprising that one can often find oneself endlessly adjusting the tone controls on one's hi-fi system in order to re-balance the relative bass and treble levels from one CD to another. Surely the frequency balances heard on many CDs were not what the mixing personnel intended, the deviations are just too great for that to be true.

8.2 Different paths to a similar result

Mastering engineers hear a very wide range of current product, day after day. They thus get a good feel for what a 'normal' balance should be, and can rapidly alert producers, musicians and recording engineers to any 'abnormality' in the recordings. Spectral balance, unconventional hardness, headphone compatibility problems, compression levels and a whole host of other idiosyncratic differences can be discussed. If deemed necessary, any problems can be rectified before any undesired characteristics travel to the factories, and on to the shops. Mastering engineers can be an island of security in an ocean of uncertainty. However, as one cannot possess their ears, so one cannot dictate which monitor systems are best for them. Their experience in making judgements on equipment which works consistently *for them*, is the combination on which their clients will be relying. They may use monitors which sound a little bright to some people, or conversely they may use softer systems. What is important is not the absolute performance of the monitors, but that they tell the mastering engineers what *they* need to know. Familiarity with the system of reference is essential, but it is worth noting here that the loudspeakers which they use for work may well *not* be the ones which they choose to use at home for their personal listening pleasure.

During the creative process of record production, different properties are frequently demanded of monitor systems, with the potential output capability requirements of the systems gradually diminishing from the initial recording, through mixing and mastering, and finally to the domestic environment. This is not absolute, as sometimes relatively small loudspeakers, or even large loudspeaker systems, are used throughout the whole process, but it *is* a general trend. The mastering process is the final chance to catch any problems before the recordings go on sale. It is also at the interface between the production process, with all of its artistic quirks, and the record buying public, most of whom just want to enjoy the music, unaware of all the 'important', 'life or death' issues which sometimes override reason in the recording studios.

From my own point of view, I like to master on two monitor systems; one of a very high degree of sonic neutrality, and one of a more typically domestic nature. The first system I use to listen for defects; the transparency, openness and revelation of details being more important to me than absolute uniformity of the frequency response. Cross referencing to the domestic system will help me to judge the overall balance of instruments. Because of the wide range of responses of domestic systems, together with the widely differing opinions of producers, engineers and musicians as to what frequency balance is optimum, the precise balance which I choose will often be based on the appropriateness to the type of music in question.

If I am listening to the final mixes of a limited interest or classical recording, then I will be much more likely to trust my own ears when listening on what *I* consider to be an appropriate monitor system, but there are commercial realities to be faced when working with popular music which is likely to sell in large quantities. Bucking the general frequency balance trends in this area may be unwise, even if the amount of low frequencies is unnaturally high, as with dance music for example. But enough of my personal views, as I am not a mastering engineer per se. However, when I do talk about mastering, I do it based on over 30 years of recording experience, and after having worked with some of the finest mastering engineers in the UK, Europe and the USA. Even now though, my golden rule is still 'Never do it alone'. I always run things past another trusted pair of ears before committing anything to release. Anyhow, let us now consider the views of a couple of front-line mastering engineers.

After completing the mix of a live classical piano concert which I had recorded, I passed the masters through the hands of Mike Brown at CTS in London, which gave me a good opportunity to discuss *his* views, so let us hear from *him*.

MB: Historically, disc cutting was always regarded as a critical stage; this had to be the case, with vinyl being a mechanical medium having its own peculiar properties. Mastering for disc remained in the hands of a few specialist engineers and studios. Even with digits, this situation should still be the case, but it isn't. Anybody can copy a tape, edit it into a final running order, and send it to a disc manufacturing plant. The plant will simply clone it to the preferred format for glass mastering, and that's it, your CD done!

Of course, some people will be happy with the result of the CDs prepared in this way, but just as many may be disappointed, often without quite knowing why. I suppose one cannot blame the plants for accepting raw DATs, but I think it's bad for the industry. One of the major roles of mastering engineers is to make sure that the CD plants get what they need to ensure that things go smoothly. CD plants rarely check incoming masters for audio faults, hence a recorded mute, for example, will faithfully appear on the pressing as part of an accurate transfer of the digital data that has been supplied. The temptation to bypass the mastering facilities (usually for financial reasons – P.N.) leaves so many customers dissatisfied, and requires real 'high-end' mastering houses to drop their rates considerably just to have any hope of getting this type of work, which in turn, limits the facilities which they can offer. By-passing the mastering houses can be seen as a cost saving, but in all too many cases, the end product suffers.

So what constitutes a good mastering facility? What exactly are the advantages of using one, and how do you know what you're getting for your money? These may seem like basic questions, but to the new generation of home studio users, these perhaps are the questions that they might like answers to. One answer to the first question is that a good mastering facility is one which gives you what you want on your finished record, regardless of what you have on your master tape (which may or may not be what you intended).

CD mastering tends to fall into one of two categories. One is straightforward transcription, though it rarely *is* totally straightforward, and the other is a more creative involvement. In the second case, a mastering engineer can perhaps offer advice on the running order, or make general suggestions about the use of an alternative mix; or even get deeply involved in reducing previously unheard noises or other problems. Oh, and of course, how does the whole thing sound; especially when compared to the generally accepted 'normal' balances.

Mastering engineers ask two important questions before starting any job: 'Is there anything wrong with it?' and 'If so, is it anything that I can fix?' To answer these questions it is necessary to have monitoring which is as 'accurate' as possible. In an ideal world of course, all EQ and compression would be done as part of the recording or mix-down, but this *isn't* an ideal world, and there are five main reasons why it may be necessary to change the sound at the mastering stage.

1 They had duff monitors in the mixing room ('It sounded OK when we mixed it, it was great last night!')
2 Something technical has gone wrong. It should be noted here that certain popular hard disc signal manipulation/editing systems are by no means as sonically accurate as they claim to be, especially with some of the earlier software. To make matters worse, a 'budget' recording set-up, of the type which would seek to avoid the cost of using a good mastering house, would also be likely to have cheap monitors, which in turn would be unable to show the sonic degradation caused by the equally cheap (in relative terms) digital work station!
3 For whatever reason, they did a duff mix. This could be down to (1) above, or that they were tired, in a hurry, using strange equipment, or a host of other reasons.
4 It could not be cut on to vinyl the way it is, e.g. there is an extremely bright hi-hat at a relatively high level towards the right hand side of the mix (which could destroy a very expensive cutting head! – P.N.)
5 The engineer and producer are, ... how shall we say this diplomatically, ... perhaps lacking the necessary experience required to produce an acceptably professional quality master. Very sadly, this is a growing problem.

In the first two cases, it may be possible to find an EQ which restores the sound to that which people *thought* that they had without too much compromise, though often, duff monitoring will have caused them to get the bass to bass-drum ratio wrong for example, so that no simple EQ will correct the problem. In all the other cases, a greater degree of compromise (and patience) will usually be necessary.

So, given all of this, what are the attributes a good mastering house should have? The two most important things you need to know in a mastering room are:

1 When you have a good tape that should be transferred flat. [Hooray!]
2 When a duff tape cannot be improved any further.

My own experience has suggested to me that the most important things which I need from my monitor system are low harmonic distortion and low coloration. Low frequency extension to 20 Hz is nice, though sadly, very few rooms used for mastering will support such frequencies. The frequency response should be smooth, though arguably, the relative amounts of LF, MF and HF are less important, as they can easily be 'learnt' by a good resident engineer, but non-linear distortions and narrow-band colorations can lead to subjective inconsistency on different types of music, and hence to difficulties in making decisions about EQ. That is to say, you end up making adjustments because they make the track sound better on the monitors you are using at the time, but on hearing the product elsewhere, you end up thinking, 'Now why did I do that?'

This would sometimes happen to me when using Tannoys in the past. It didn't matter how carefully they were set up, they would still end up fooling me. Later, when I started using ATCs, I found they were much more informative and would 'talk' to me when I touched the EQ, saying 'What have you done that for? It sounds different, but it is no better, is it?' I began to hear lots of things that were wrong, but which it was far too late to try and fix. And that is important: 'To EQ, or not to EQ?', that is the question.

Different mastering engineers may differ about their choice of monitors though. (In fact, a colleague of Mike's used Tannoys regularly, from choice.) Of course, we *all* have different ways of listening to and interpreting what we hear. As a mastering engineer, you can listen to a song half a dozen times, listening for different effects of EQ, compression, noises and general problems, yet still not know a single line of the lyrics; and may even have to ask what the song is called. Mastering engineers cannot be expected to like every piece of music which they are asked to assess, so despite the need to listen to the whole in a subjective way as an overall assessment, the prime function of a mastering engineer is to be as objective as possible. Even so, I have often thought it surprising that experienced professionals can hold such widely differing opinions about monitoring. Engineers are often strongly prejudiced in favour of the familiar, assuming that monitors that are unfamiliar are 'wrong', simply *because* they are unfamiliar. In fact monitors are *always* wrong – there is no such thing as perfection, you just have to learn the imperfections.

I currently use ATC SCM100s on Recording Architecture stands. They are set flat, and actually sound it. They don't go down to 20 Hz as I would like, but the stands make the LF much better, and I feel that I can make good, consistent and objective judgements when using them. I'm not completely convinced, however, that conventional cone-type speakers are the be-all and end-all. I have always been impressed by the Quad Electrostatics in their various forms, and feel that it is a shame that this kind of loudspeaker technology isn't given more serious attention, which brings me to the point which *you* [P.N.] have made many times, both verbally and in the recording press: 'Are the Quad Electrostatics *so* natural, that they are not representative of 99.99 per cent of loudspeakers as a whole? Not that they are too good to be *true*, but that they are too *true* to be representative?'

The problem with monitors is just one of the reasons why I feel it is as important as ever, even in the wonderful New Digital Age, to have recordings properly mastered by experienced mastering engineers, rather than just sending a 'do-it-yourself' job to the factory. Unfortunately this is a difficult and abstract point to argue, and the people who should most *need* the advice usually find it the hardest to understand. This is a result of the current free-for-all created by ever falling 'semi-professional' equipment prices, but I don't think that it is a good thing for the reputation of the industry as a whole.

Well, such are some of the views of Mike Brown, all of which I can perfectly well understand and respect. However, before proceeding to the views of another mastering engineer, I thought that it would be good to look from the viewpoint of a well-respected and experienced recording engineer. I decided to pick the first appropriate person who I bumped into, as that would reduce any possibility of seeking out like-minded persons via any chain of recom-mendations.

Soon after speaking to Mike Brown, I found myself in Lisbon, to witness the first trial sessions in a studio which I had just completed for Rui Veloso. (See Figures 6.3 and 6.4.) His last album but one had by then gone eight times platinum in Portugal, breaking all previous national sales records. At the time when the studio was completed, he was playing in the Trump casino in Atlantic City in the USA, and had recently been invited to play again with BB King on his new album. For his next album, Rui had chosen to work with the Irish producer/engineer Rafe McKenna, whose other international suc-cesses have included work with Bad Company (platinum album in USA), Richard Marx, UB40, Six Was Nine, Wet Wet Wet, Foreigner, Paul McCartney, Depeche Mode, Thomas Dolby, Giant and a whole host of other well-known artistes. He is reputed for being able to relate equally to rock, contemporary, R&B, classical and 'left-field' music.

I had known Rafe from way-back, when I had completed the building of The Townhouse in London in 1978. Barbara Jeffries and I had concluded that the only way in which we could launch another super-studio in the then current economic climate was to staff it with the finest personnel, from top to bottom. I had been impressed by Rafe when he was an assistant engineer at The Music Centre (now CTS) in Wembley, London, which I often used for mixing *and* mastering recordings which I was doing for Virgin. We offered him a job as assistant engineer at The Townhouse, but The Music Centre had just promoted him to recording engineer, and he was reluctant to 'downgrade' again, though on similar pay and perhaps better conditions. He declined the offer, and the job was snapped up by Hugh Padgham, who was also on our short list of potential 'super assistants'. In 1978, Rafe considered Virgin to be a company of little significance, but laughs about his error, now.

I had arrived at Rui Veloso's studio the same day that Rafe had flown in from London for the test sessions, so an excellent opportunity arose to hear his views. Surely a person of Rafe McKenna's experience and track record would be able to make up his own mind about when his masters were as he wanted them to be, so would a person of his calibre *need* to consult a mas-tering house before sending the tapes to the factory?

R.M.: Mastering is an important part of making any record. Whilst there are many engineers like myself who have the *ability* to do this job, it's generally passed to a mastering engineer who has particular expertise in this area. In fact, I always use a mastering facility when at all possible. My only real unease about any of the CDs which I have recorded is when occasionally, work that I have done in other countries is mastered either without me, or without using a reputable mastering facility.

I usually use my own monitors for recording and mixing but if the monitoring in a studio is good, it's an added bonus. During the record-ing and mixing stages, I most definitely am the person who chooses the monitoring which will be used, but when mastering, I use the facilities because of the people I know and trust. I let them use the monitors *they* know because I trust *their* ears and *their* general competence. I know that by using a trusted facility I can be fully confident that the EQd and PQd production master has been checked for any errors or glitches before it goes to the factory for the glass/metal master. I always try to use top mastering facilities, but the precise choices are based on the ones which consistently work for me. When I've worked abroad and I've not been involved in the mastering, I *have* had a few surprises with the finished CD. Also I've found sometimes the finished CD sounds harsher or more steely sounding than how I remembered it.

As regards general frequency balance, I personally don't make any compromises to current trends or fashions. The end result I go for is what *I* feel is right for the project, in conjunction with the producer and artiste I'm working with at the time. This overall approach has worked well for me, and I see no reason to change it, but it is often only in the mastering room that the final satisfaction with the work is confirmed. I do occasionally work for non-UK record companies who are tempted to bypass the mastering, on grounds of cost, but relative to the overall album budget it is such a small proportion that it seems very unwise to try to cut costs here. For me it is a *very* important final step in the record production process.

I hope that it can now clearly be seen from Rafe's comments that despite the temptation for many people to cut costs by by-passing the mastering phase, the facilities which the mastering houses offer are usually excellent value for money. They ensure that the maximum potential is realised from the mixes which leave the studio, and also help to ensure that the record buying public are not short-changed. The sheer cost and difficulty of the operation of disc-cutting systems for vinyl ensured that everything released passed through a professional mastering process. If nonsenses *were* released, then it was usually deliberate, and not through ignorance (though ignorance sometimes *could* be the reason for the deliberate release of nonsense).

The ever reducing real cost of the whole recording chain has allowed so many people of inadequate training (and who only *think* that they know enough to do the job well) to send masters directly to the factories via the record companies, the less responsible of whom have been only too glad to reduce the overall cost of production. The effect has been a very hit-and-miss approach in terms of the quality of the end product. It is tempting, very tempting, to allow yourself to be convinced that your master is 'the one' but the closeness of involvement, especially over a relatively long period of time, can draw many people into a virtual reality. Under this influence, it is often the *real* world which becomes seen as unreal, and a deep belief can be cultured as to the inherent 'rightness' of the masters. Even some of the most experienced professionals can suffer from this, but at least they are *aware* of it, so use trusted mastering engineers to help to keep their feet on the ground. One of the greatest lessons of professionalism is to know your limitations, and the very top people are never afraid or embarrassed to ask for help.

I do not know of any serious mastering engineers who would work solely by reference to a pair of NS10s or the like, as they need a much greater degree of linearity and resolution of detail to be able to perform their critical job; but within the higher performance range of monitors, the choice must be down to the preference of each engineer. On the other hand, so many of the people who bypass the mastering houses *do* assess their work on totally inadequate loudspeakers, which the experience of the mastering engineers suggests that you simply cannot do if the true sound of the mix is to be heard. If non-linear distortions and any harshness from the build-up of processing and effects are to be heard to their true extent, then they cannot be allowed to be masked by the inadequacies of most inexpensive loudspeakers.

Mastering engineers need clear monitors to be able to do their job, but mastering is more than just monitoring, mastering is about judgement and objectivity, with a touch of creative and subjective advice thrown in. When I write a chapter such as this, it will pass through the hands of a sub-editor before publication. Technical problems aside, such as typographical errors, I cannot recall ever having had my work reduced in its potency at the hands of a good sub-editor, and on many occasions, there have been significant improvements made to the clarity of the writing. Anyhow, a good sub-editor will always be available to discuss any changes, and if I really feel strongly about a point, then probably my views will be respected.

Good mastering engineers can be thought of in the same way as good publishing sub-editors. Their skill and judgements are sharpened by the sheer volume and variety of material which passes through their hands: they get a good feel for what is right, and what is balanced. In both cases, the better the *standard* of the material that passes through their hands, the more aware they will be of the true state of the art, and of what is realistically achievable. If you pass the critical tests of their eyes and ears, you can be pretty sure that you will not be letting yourself down when the end product is available for public consumption. They are critics who are working for *you. No* degree of experience puts *anybody* beyond the odd embarrassing blunder, and so it is better to receive a timely word in your ear, rather than to release something which could prove to be a very public embarrassment.

It seems from what Rafe McKenna has said, that there is *no* level of experience that out-grows the need for reference to a good mastering house. So, to close, let us get the views of another very experienced mastering engineer, Gordon Vicary. At the time of publication, Gordon had moved on to Soundmasters, but for almost 20 years he had been working at The Townhouse, in London, in whose mastering/cutting rooms the staff must have accumulated around 100 years of combined mastering experience. The Townhouse 'Sonic Solutions' room is shown in Figure 8.1. Gordon began work in the mono-cutting room at Pye, London, in late 1970, and in those days, the restrictions were so great that it was sometimes like working in a strait-jacket.

G.V.: Things have changed somewhat since then. With the majority of tapes that we had to cut, it was generally considered that we could get through an album and a single before lunch, then two albums between lunch and 6p.m. These days, 15 hours is quite normal for one album, and even that is pushing it if you have to make separate production masters for vinyl, cassette and CD.

The majority of the jobs which I do are attended by the engineer or producer from the mixing. Since I moved to The Townhouse in the early 1980s (from Utopia) it has always had some of the best designed and best equipped rooms in London, so we rarely got any back-catalogue re-cuts, because the rooms were always in such demand for high profile work. That has been a huge benefit in terms of experience, as some of

Figure 8.1 The 'Sonic Solutions' room at The Townhouse, London, UK

the greatest producers and engineers have passed through, with some of the finest recordings of the very top artistes. This helps to set a good reference standard with which to work.

The only way to really appreciate what you are hearing is via a monitoring system capable of reproducing the mix faithfully, with the absolute minimum of coloration – from the loudspeakers *or* the room. Having the luxury of the facilities which I have enjoyed for the past 20 years, I have on many occasions been able to point out something in the mix which the client had never previously heard. We currently use PMC-BB5 monitors, with Bryston amplifiers, as the main system. To us, the PMCs are the most natural sounding of all the ones which we tried, and we were very pleased with their performance at all monitoring levels. However, I still like to use both large *and* small monitors. I tend to EQ on the large ones, then check what I have done against the smaller ones, then refer alternately to each until a generally satisfactory result is produced.

In many cases though, it is not so much what you hear, but what you are listening for. The experience gained in the early years of my career gave me a feel for sounds which many non-disc-cutting engineers may never be able to experience. You had to listen to the music when cutting on the old lathes, but you also had to listen *through* the music, to try to detect anything in the overall sound which suggested potential problems. Excessive high frequencies, for example, could easily blow a very expensive cutter head. It turns out that so many of the things which actually *were* problems to transfer to vinyl are things which do not do the overall balance, smoothness, or stereo imaging any good via any other medium. Certain sound characteristics ring warning bells in my brain, which was highly sensitised when having to work on the older disc-cutting systems. Even though some of these smaller things are no longer problems in themselves, it can still be the case that the removal of any minor problems, and sometimes even hidden ones, can be the difference between a good recording and a great one.

The wider range of experience of different sounds and different ideas which a busy mastering engineer builds up is also very useful when compiling albums which consist of recordings from different studios, and which were perhaps recorded by different engineers and producers. For me it is still a real joy to get a consistent and homogeneous sounding album from a 'difficult' assortment of songs; and even more so to be able to equalise at will to achieve the best overall result for CD, rather than having to filter down to a lowest common denominator, which sometimes used to be the case with vinyl. Not as much importance is now given to recording for vinyl, but we still have to take it into serious consideration as the medium still exists, and clients would complain if a subsequent transfer didn't sound good, even if they were not considering it too seriously themselves at the time of the original mastering. We don't have any regular clients who are still fanatical about vinyl, but we do get the occasional client for whom vinyl is still a very important medium.

As far as people doing their own mastering is concerned, well, once again, remember that because mastering engineers are often working with the very top levels of musical production, they have a ready reference to what *can* be achieved, so are very aware when something needs to be worked on to bring it up to standard. There always has been, and still are, a number of really good recording engineers who seem to produce great sounding mixes, almost irrespective of where or with what they were recorded. On the other hand, it must require every bit of their skill and experience when working on cheaper equipment. In general though, many of the recordings done by lesser mortals on cheaper equipment often do not seem to have the full sonic spectrum. These people often work with cheap monitoring, so unless they have years of experience, they will not be aware of what is missing. In many cases, the work is musically excellent and well balanced, so it is a pity when small points let it down. They may over-use low frequency filters, supposedly to 'clean-up' the recording, in the belief that these filters are not affecting the mix because they will not *hear* any effect on cheap monitors. What often happens when we receive such tapes is that it soon becomes apparent exactly what is missing. Frequencies that would need boosting to match the sound of high budget projects are either just not on the tape, or contain previously unheard problems, which restrict the amount of boost that can be used before they would become too plainly apparent.

From a personal point of view, it is very rewarding when something can be salvaged or significantly improved by all of my efforts, yet at the other end of the scale, it is also rewarding to take part in the transfer of superb masters which need nothing doing to them whatsoever. I suppose that you may wonder why such good quality masters pass through my hands, but I suppose that the answer to that is just for one last check to make sure that all is well. It provides reassurance to the recording personnel, who often, at the end of a long and difficult job, are too tired, too stressed and too close to the proceedings to be absolutely sure that they are actually hearing what they think that they are hearing. For me, there is satisfaction from almost all of my work, whether it is salvaging a disaster from the dustbin, or just giving the final nod of approval to the work of a megastar. Perhaps my only real disappointment comes from hearing releases which good mastering engineers could have significantly improved for a relatively small amount of money, but which the people involved saw as an unnecessary expense. It is frustrating to think 'I could have saved that for no more than the cost of a couple of hours of studio time'. This is especially so when the bypassing of the mastering studios is only out of ignorance.

8.3 Summary

Mike Brown, Rafe McKenna and Gordon Vicary each have between 20 and 30 years of professional experience behind them, and the number of recordings in which they have been involved is truly beyond counting. All seem to agree that the mastering houses have a great deal to offer in many different ways. They can squeeze the last ounce of potential from a less than wonderful recording, they can provide a last minute check of a 'perfect' recording (to help everybody's peace of mind), they can tidy up small problems, and perhaps most importantly, they can provide a sympathetic, understanding, but nonetheless objective pair of ears, to listen in a more detached way than can the recording personnel. They seem to agree on low coloration, low distortion monitoring conditions as a prime requisite, with absolute amplitude response flatness being nice if you can get it, but perhaps being of less overall importance, as they are listening for details but comparing tonal balance to known standards which they refer to daily. In many ways, good mastering engineers are 'The professionals' professionals', whose true worth is frequently underestimated by the general public, and indeed, by perhaps too great a proportion of the music recording industry itself.

One reason for writing at such length about mastering in a book about project studios is to help to highlight some of the aspects of monitoring which cannot always be readily appreciated from working in limited circumstances, or from reading advertisements. A friend recently returned from an AES exhibition, somewhat confused by the fact that every loudspeaker salesperson seemed to be claiming that *their* products were the best. In effect, it was impossible to know who to believe. On that point, perhaps it is best *not* to listen to loudspeaker company representatives, but to pay more attention to the views of experienced users and, more importantly, one's own ears. Hopefully, the views expressed in this chapter will have given some insight into not only the degree of personal preference, which is inevitable, but *why* some of the people involved lean in the directions that they do. However, it is all ultimately down to experience, and not what anybody else tells you.

On the point of the hi-fi/studio monitor paradox, the mastering engineers are the people who above all must work at the interface of the two camps. It is their special expertise which is relied upon to ensure the best transition of the recorded material from the studios into peoples' homes. I set about writing this with a purely open mind, but was not too surprised to find that none of the people mentioned by name in this chapter had chosen the same loud-speakers. I know that Rafe McKenna is, at the time of writing, choosing to use KRK loudspeakers as his mid/close field monitors.

When I am building studios, and when the choice is down to me, I do have preferences as to which main monitor systems to use. Indeed I would, as I specifically designed a range for Reflexion Arts to suit my own room con-cepts, *and* my own ideas of what I, and most of my clients, feel that it is important for us to hear. Essentially, that is a sound which is as sonically neutral as we can currently achieve, as I believe that the very term 'monitor-ing' suggests something of a more objective reference. However, for some years now, I have been reluctant to get too involved in the individual choices which lead to the purchase of the close-field monitor systems. In reality, it is an area in which the choices turn out to be subjective in the extreme.

I still feel that it is in most cases necessary to have a good, sonically trans-parent, non-flattering, wide frequency range main monitor system for general reference, but exactly which small monitors are 'the best' in helping to achieve a desired overall musical mix, remains highly personal. The analogy with the choice of tennis racquets probably applies quite strongly, and cham-pions of neither recording *nor* tennis will achieve their results with bad equip-ment. Ultimately, though, it is neither the racquets nor the loudspeakers, as such, which make the champions; it is the experience of their professional users. In the end, and beyond a certain point, nothing can substitute for expe-rience, and if nothing else, good mastering houses are the places where so much of that very necessary experience is to be found.

Chapter 9

Some further items for consideration

The 1980s saw an enormous change in the approach to the design of recording studios. Where the mid 1970s saw studios being constructed mainly for experienced companies, the mid to late 1980s saw a great upsurge in the number of studios built for 'first time' operators. The mid 1970s assumptions that the client largely knew what was required could no longer be taken for granted. Nowadays, in many instances, a prospective studio owner begins discussions armed with a list of equipment *names* which he or she considered indispensable, … and very little else! The lack of experience in the initial set-up leads to many owners of completed studios failing to realise what it takes to operate a studio, even semi-commercially, then panicking, and charging ridiculously low amounts of money for their studio time in desperate attempts to survive. We now seem to have many people in the profession with far more money than experience. This applies to studio owners and users equally. Inexperienced clientele often make totally unreasonable or wholly impossible demands upon studio owners, who, due to intense competition or lack of experience, end up by bending over backwards to the point of obsequiousness. This only encourages more unreasonable demands to be made, which leads to a situation akin to countries giving in to or making deals with terrorists, … that is, it spawns more of the same activities.

When designing studios for the twenty-first century, a greater understanding is required of the philosophies underlying the modern usage of studios. Recording studios in general are not being used in anything like the same way that they were being used 10 or 15 years ago. The problems encountered are rarely the direct fault of the studios concerned, but are more likely to be a result of either the studios being inappropriate for the purpose for which they are being used, or the inappropriate techniques which are being applied, given the systems available. This point is discussed, further, in Chapter 14.

9.1 Reference standards

Studios used to be in the vanguard of the pursuit of excellence in recording. Now, many seem more to the forefront of the pursuit of turnover and mediocrity. It still seems very strange to me that people can spend $400,000, or more, on a studio, and proceed to judge the entire system through a pair of loudspeakers costing under $400 to the studio. To put it a different way, everything is measured by a piece of equipment – a pair of loudspeakers – which may represent only one-thousandth of the budget. It is therefore little wonder that consoles, and other pieces of equipment, are often on the market for a considerable period of time before many people become aware of their sonic deficiencies. Such subtleties are just not apparent without a high quality system of assessment – good monitors, and a good room!

Decisions are being made on very scant information. For example, I was asked by a studio owner for my opinion on the decision between two sec-ondhand consoles, 'A' and 'B', for his second studio. After due discussions on the subject, he spoke to me a few days later and said that he had been speaking to a member of a well-known band who assured him that the choice should be console 'A', then listed a whole series of pro-'A' points. I queried a few details and the studio owner asked me to speak directly to the pro-'A' pundit, and I duly obliged. Further questioning revealed that the person in question had never even been *into* a studio with a 'B' console, let alone worked on one. He had not been asked about his opinion of 'A' consoles, but specifically on the choice *between* consoles of types 'A' and 'B'. His answer was of course invalid with respect to the question asked. His reply probably would, in the absence of the more detailed questioning, have swung the studio owner in favour of the 'A' console, which may or may not have been the most suitable choice. Studios built to such arbitrary criteria can never be expected to advance the process of recording. Only the continuing pursuit of excellence can move the recording industry forward. It should be noted, though, that the pursuit of excellence may be more greatly rewarded by an investment in knowledge, than purely in money. I strongly believe that the lobby advocat-ing 'compromise to the market' will only lead to a general decline.

And the standards by which many people judge all this? More and more by what things sound like on somebody's Walkman, or in a car. Imagine trying to measure how hot we can make a crucible of molten silver, with a ther-mometer which has 100 °C as the top end of its scale. We can deduce only two things: firstly that the silver is above 100 °C, and secondly that it is also above its melting point. Just how hot it is we do not know. Similarly with recordings, unless we have high quality, wide range monitoring as a reference (and by all means keep some lesser quality close-field monitors to reference *down*), we cannot judge just how good a recording we can achieve.

9.2 Design integration

It is time to begin looking into the erosion of recording practices and the underlying tendencies in this very much expanding industry. This ever-chang-ing state makes it difficult, very difficult, to achieve satisfactory results. A brief such as 'I want to use the monitoring from studio A, the back wall from studio B, the front wall covering of studio C, the style and shape of studio D', and so forth, is not a viable design philosophy. It is more akin to the con-tention that a committee, attempting to design a race-horse, will be likely to produce a camel. An integrated approach is essential to achieve the best results. The front wall covering of studio C may well be totally incompatible with the back wall design of studio B. The argument usually put forward in defence of this is 'Well, in studio F, they didn't have a designer and it sounds great; so all of this designer stuff is a load of hype!' Granted, studio F may well sound great, but it could be a fluke. Footballers have scored great goals in cup finals from what they had intended to be crosses, but no self-respecting football coach would advocate the mis-placed cross as a viable technique for scoring goals. There is only one company which springs to mind which has succeeded in using the technique of multiple component source design as described above. Rolls-Royce cars are excellent, though rarely innovative. They chose, in years past, the best aspects of other manufacturers' products then carefully refined them and moulded them into their own philosophy. I believe that they paid royalties to Oldsmobile/Cadillac for engine design, General Motors/Borg-Warner for gearbox design, Hispano-Suiza for braking system/suspension design, and also to several other manufacturers. Rolls-Royce show that it can be done well. They also show something else – making such a philosophy work is a *very* expensive process! Attempting the same thing with random parts from a scrap-yard is likely to lead to disaster. Project studios are not usually built on huge budgets, so the Rolls-Royce results are not likely to be achieved by this 'mix and match' approach.

9.3 Flawed concepts

Life in a recording studio is becoming an ever-more complicated affair. If the number of variables is now reaching alarming proportions, maybe this explosion to some degree explains why people seem to be turning to over-simplistic theories and expectations in order to find some secure sanctuary in the chaos. Like the burying of heads in the sand, however, over-simplistic expectations do not remove the source of the problems. The *real* problems which confront us are dictated by the laws of physics. The natural laws are very powerful, and no amount of arguing will make then go away.

9.4 Finding a path

With a certain amount of rationale – opposed to pseudo-scientific dogma – we can begin to unravel some of the confusion. More so than ever before, we have a very wide spread of shapes, sizes and designs of studios, and a more competitive and ever-more demanding clientele. We have a lower proportion of staff trained to the degree of previous years, and for much of the time, we are recording sounds with no natural references. With computer generated sound we cannot 'stick our ear to the instrument' to compare live, with recorded sound.

For anybody without a great deal of experience, all of the above points breed insecurity, and it is this insecurity which now consumes so much time in the conception of design philosophies for new studios. If this is the way in which the industry persists on progressing, then so be it. It would be comparable to swimming up a waterfall to try to change the flow of things now; the momentum is too great! Surely though, more understanding of the problems involved can do no harm.

False syllogisms are frequently applied, on the lines of 'Cats have tails;

Fido (a supposedly common name for dogs) has a tail; therefore, Fido is a cat. Likewise, statements about loudspeakers can draw similar false, or at least unsound, conclusions. Here, however, even the input statements may be of dubious relevance to the real problem. One hears 'loudspeaker of type "X" sound great in room "Y" with all my record collection, therefore I shall always use them to judge the correctness of a mix (implying in whatever surroundings).' In terms of reference monitoring, the question is, or should be, do they sound *accurate*? Even if all the recordings do sound great, have all the recordings been checked by playing them in the rooms in which they were originally mixed, in order to hear how they were *intended* to sound? Furthermore, were these checks carried out with the original producers present to confirm that the sound is still as was originally mixed? That the collection sounds great to one person on the 'X' loudspeakers in room 'Y' is in itself arbitrary. Rooms have an enormous effect on the character of sound, and the positioning of loudspeakers in any room is of great significance. The characteristics of the room *and* the loudspeaker can enhance and detract from each other, producing sounds which bear little relationship to either the speaker or room in different combinations. We can also become accustomed to a certain sound profile which leads us to confuse accuracy with familiarity, our brains have an amazing capacity for acclimatising to, and compensating for, deficiencies which are regularly encountered. Unless care is taken, we can begin to look for sound similar to those with which we are accustomed, feeling certain that such sounds are 'right'. Contrast this situation to the build-up of experience in mastering rooms, where reference back to the original intention and sounds of a mix are usually available via the presence of members of the original recording personnel.

9.5 What is right?

'Right' is in itself an extremely arbitrary word. We have neither control rooms nor loudspeakers that are absolutely accurate in either objective or even subjective terms; at least not for a wide range of music. Given the present state of technology, I see little prospect for perfect monitoring in the foreseeable future. We must accept that we are dealing with degrees of compromise – not absolutes; however, this should not prevent us from *seeking* the absolutes. Loudspeakers, like microphones, are pressure transducers. They have one, and *only* one, plane of diaphragm movement, whereas a natural instrument emits vibrations in three dimensions. In terms of positioning, we create 'phantom' stereo images from two relatively small source areas on the left and right. A group of musicians spread in front of a listener produce individual, integrated, three-dimensional sound sources in their own right. Considering just these facts shows us that conventional recording systems are far from ideal or perfect. Given that anechoic chambers may produce grossly unrepresentative listening environments, at least in terms of domestic acoustics, they may well still be the only definitive answer to accuracy. We must accept that all control rooms will be imperfect and will have some sonic characteristics which will colour the sounds to which we are listening. The colourations due to each room will be different, and the overall perception of the sound will be different. These facts are fundamental to the concepts described in Chapter 11, 'The Small Room Problem', which outlines an approach which seeks to minimise these irregularities.

Whilst continuing to strive for perfection, unless we accept our current shortfalls, we will be using entirely the wrong yardsticks to judge workable performance. Just like rooms, the loudspeakers themselves have individual timbres which have a bearing on the overall sound. Loudspeaker manufacturers also tend to have their own 'in house' sounds, despite perhaps showing great similarities in the measured performances of their products. On numerous occasions, I have been called to studios because a snare or tom-tom appeared to change pitch when switched from one loudspeaker system to another. All drums have resonances; if they did not, there would be no character whatsoever to the sound. Music as a whole is essentially a collection of controlled resonances. Percussion, however, in untuned form, is different from other musical instrument groups in that the resonances do not appear in such controlled, harmonic structures. Percussion is more akin to a blast of noise with random resonant effects, and indeed this was the means by which early drum-machines operated. Just as with a tuned instrument, percussion tends to adopt the pitch of the most enhanced resonance. These random resonances are frequently found in quite evenly balanced quantities, but different loudspeakers, having marginally different responses, may well emphasise or reduce some of the resonances of the drums. These relatively minor discrepancies can be sufficient to tip the emphasis from a major resonance when played on one loudspeakers system, to a secondary resonance when played on an alternative system. In consequence, the apparent pitch of the drum can be perceived to have changed. This is not unusual or disastrous, as we have already concluded that *all* loudspeaker systems are flawed. Ninety-nine times out of one hundred, when the drum is mixed in with the remaining instruments, the problem is effectively masked. I have seen numerous studio personnel waste hours, or even days, and a great deal of money, trying to solve a problem in a room, or in a monitoring system, when a visiting engineer has complained about the above discrepancy in sound between large and small monitor systems. Even more alarming is the virtually unquestioning faith, by some people, in certain close-field loudspeakers, and the belief that they are accurate in all environments and on all forms of music.

In yet another studio to which I was once summoned, a resonance was apparent on one particular, synthesised bass note. After many, many tests, it was shown that the sound in question triggered a resonance in the above mentioned small loudspeaker. It was ultimately shown that the problem of resonance existed only on those particular loudspeakers in that particular position in that room. It was subsequently found to exist to a far lesser degree on loudspeakers of a similar model in other rooms. Substitution of many other loudspeaker types in the original room failed to show any sign of the problem. The studio owner logically believed that the combination of that pair of loudspeakers and the room characteristics combined to produce an unfortunate, but rare effect. Indeed, I myself felt that the only reasonable, logical conclusion to be drawn was that the type of loudspeaker in question was prone to the excitation of a particular resonance in the room when triggered by certain, difficult, and soloed sounds. Unfortunately, the studio manager was faced with the undesirable task of convincing the producer of this conclusion. In actual fact, the attitude of the producer was that he had never come across the problem before, and knew these loudspeakers well. His suggestion to the studio manager was that if the problem was not apparent on any other loud-speakers in that room, then it merely showed that the other speakers were not as revealing. In effect, the producer's statement was tantamount to the attitude that these inexpensive loudspeakers were right, under all circumstances, and that every other loudspeaker in the entire world was wrong. Unfortunately, these attitudes are all too common, despite being totally erroneous. So, how does a studio owner deal with such a situation? Certainly not by logic; and if this industry persists in making its fundamental measurements with elastic tape-measures then it can only lead to confusion and ultimately to retrogres-sion. It would seem that there *is* no simple answer for the unfortunate studio owners.

9.6 Experience or insecurity

With the rapid expansion of the recording industry and the alarming rate at which new equipment becomes available, I am convinced that insecurity now plays a large part in the making of many decisions. Once again the psychol-ogy of providing a secure environment for visiting engineers and producers becomes as much a part of the studio's service as the provision of the equip-ment and other facilities. Many producers and engineers have their own favourite studios. This choice is often based on their feeling of security rather than the other merits of the studio itself. They know that turning a certain knob in that studio produces a certain effect. When confronted with a differ-ent studio, these people will, almost *ad nauseam*, compare every detail of the studio in which they are working to the studio with which they are most familiar. 'This doesn't sound like that in the studio I usually work in' is the sort of comment heard. 'We use a different type of patch cord' … 'We've got the effects rack in a different place' …' When I want this sound, in my usual studio I just turn this equaliser knob, but with these equalisers I can't find the sound' … and on and on it goes. I often wonder why they do not return to the studio where they came from if they dislike so many items of the new studio. It is usually insecurity that causes them to question everything, which is usually due to a lack of experience. There are certain timid car drivers who are very reluctant to drive any car other than their own. Should the indicator control be on the opposite side of the steering column, then that could confuse them totally. To an experienced driver, however, the differences from one car to another are taken as a matter of course. As long as the fundamentals are correct, in experienced hands, the car will travel safely from A to B, within the limits of its performance, and the journey will be completed without fuss. If only people would consider recording studios as tools to achieve specific end results, rather than being the centre of the entire universe during the time of their recording session. Psychologically, being entombed in a studio, often for months on end, can result in a distorted sense of perspective. I become aware of a whole year 'missing' from my life in the mid 1970s whenever I see television programmes featuring that year. I spent virtually the entire year cocooned in a studio working on one album. The album was very successful, and I still think that the effort was worth it. However, what was sig-nificant upon my re-introduction to society in general was that my efforts played an exceptionally small part indeed in the greater scheme of things. After being fed eleven months of record company hype about the importance of my task, it came as something of a surprise to find that the man two doors lower down the road was totally disinterested in the entire proceedings. Proportions were thus absurdly distorted. What is more, this was at a time when the recording industry was still a good deal more structured and orderly than it is today. At least we had been trained and prepared for the pressure of the work.

Lack of knowledge and training often lead to situations where people do not know the difference between what they should know, and what they cannot reasonably be expected to know. In an industry where people are frightened of making mistakes for fear of seeming to be incompetent (though often only in the eyes of the even less competent), it is no wonder that insecurity abounds. Everybody appears to feel that they need to have an opinion on, or a working knowledge of, everything that exists. With the rate of proliferation of new equipment, just keeping up with the details of every new product would be a full time job in itself. It is no shame to not be familiar with some equipment, it is, in fact, quite normal. Yet somehow, to answer 'I haven't seen that' is taken by many to be an indication of not being up to date, and, by implication, lacking in the required knowledge to do a good recording. Once again, this in itself causes problems. There is no shame in admitting when ones does not know something. Only to the ignorant will this appear to be a failure in the points-scoring game. To those who *do* know, it will be a sign of honesty and a willingness to learn. Honesty is fundamental to genuine progress. Having said this, I do appreciate the difficulties when one works in a highly competitive business in which the clients, which one needs to survive, are exhibiting *their* lack of knowledge, but who still manage to convince so many people that they know so much.

9.7 Effects of effects

There would seem to be a great reluctance in some engineers in small studios to accept the fact that they can greatly affect the performance of a room by the quantity and positioning of equipment which they bring into the control room. In the past, control rooms were designed to house the recording equipment and personnel, and the studios were designed for the musicians and instruments. Nowadays, control rooms house everybody and everything, except for the empty flight cases, which are now often stored in the studio areas. Resonances in the metal panels of the equipment, along with cavities created by the intrusive equipment, often positioned in resonant squares, can contribute to significant changes in the room acoustics between the full and empty states. I have actually witnessed situations in which engineers have surrounded themselves with equipment, creating reflective and resonant 'boxes'. This has completely changed the usual monitoring conditions to an absurd degree, yet the engineers responsible have completely failed to be aware of their error.

9.8 Further compromises

The nibbling at the edges of professional standards is insidious and seemingly relentless. In control room design I am frequently asked questions such as 'What would happen if we filled that bass trap with effects?', 'What happens if we leave the wall straight?', 'What happens if we increase the amount of glass at the side of the room?', and other similar questions. Taken individually, and frequently out of context, the answer to each question may be 'nothing significant'. The cumulative effect of the end result, however, if *all* the changes were put into effect, may well be disastrous to the performance of the room as a whole. We can end up with a room having no significant acoustical control. This amounts to having little more than a large, decorated room with some equipment in it. When things fail to sound as expected, the usual answer now seems to be to blame the large monitors, then proclaim with great apparent wisdom that the close-field monitors are more accurate anyway. The fact that, in so many cases, the large monitors are being asked to drive a seriously compromised set of room acoustics is lost on them. The close-field monitors are probably *not* more inherently accurate, but they get away with less criticism because they are attempting to do less. They are driving a smaller listening area with a more limited range of frequencies.

9.8.1 Limited range monitoring – the dangers

Increasing the 'certainty' of the mix by reducing the frequency range being examined is not a wise philosophy. Particularly in the realms of the lower octaves and soloing, many close-field systems are sadly lacking. This is not the fault of the manufacturers, but is a result of incorrect use. Low frequencies are frequently incorrectly balanced when over-reliance is placed on obtaining the sound via a close-field system which begins rolling off around 70 or 80 Hz. When soloing instruments, especially from computer sources, the peak to average ratio of the sound is often very great. Whilst being in no danger of exceeding the thermal power rating of the loudspeakers' voice coils, the transient peaks can easily exceed the loudspeakers' maximum rated diaphragm excursions. That is, they instantaneously move a greater distance on the attack of a note than that for which they were designed, producing distortion which colours the sound. This is somewhat ironic because these distortions disappear at lower levels and explain why the effects noted when soloing an instrument are frequently inaudible when the instrument is set back into the complete mix. In the mix the average power level rises abruptly, and hence so does the apparent overall loudness with relation to the peaks. On full-range large monitors, these problems are not apparent, which explains the need for a very high transient handling capacity of well-designed, large studio monitors. Close-field systems are complementary to, and not replacements for, large conventional systems.

Modern recordings possess an unprecedented inconsistency in the level and the tonal responses of the lower octaves when reproduced on full range systems. This is not surprising when these frequencies are *not* being adequately monitored in the control rooms. They are being left to run wild, and, consequently, you get whatever you get. The removal of acoustic control factors in the control room, the influx of unspecified and random equipment and instruments *into* the control room, the moving away from the design specified listening area of the room, the subsequent loss of confidence in the main monitors, the resort to close-field monitors ... it is a slippery slope.

9.8.2 Training

Many times, I am asked to build studios on the strengths and reputation of those which I have built previously. These potential studio owners expect that by giving them the basics of a studio similar to one with an excellent reputation, then they themselves will soon acquire the same success. Equally, they could buy a Formula-One Grand Prix racing car, but could they hope to win a race next season? They would be fools if they even thought that they had a chance. They would be forgetting, of course, the extreme levels of specialised skills possessed by the professional drivers. In their hey-day, most of the 'name' studios had large maintenance staffs to keep the equipment in top condition. The engineering staff had long and enviable reputations and the assistant engineers were brought up under 'those who knew'. There is more to a top studio that the studio itself. There is the skill, and I mean true skill, and the experience of their staff. It takes a great deal of time, patience, skill and money to weld together a highly efficient and respected combination of studio and staff. With studios maximising their investments in equipment whilst being forced to reduce the inherent level of staffing, it seems likely that, in the short term at least, we are not going to see the conditions to promote the development of many of these facilities into top class studios. To this day, I would still back a highly trained staff on poor equipment to outperform the finest equipment in the hands of inexperienced staff.

9.9 Specialisation of personnel

Another aspect of staffing is workload. With the enormous increase in the complexity of the electronic instruments or sound sources, the engineers are being called upon to work far more closely with the musicians in the programming stages. It has been said that an engineer should be no more required to programme an instrument, than to tune a guitar or piano. The reality, however, is that engineers do get drawn into this, which is not entirely surprising. In order to engineer effectively, a rapport with the musicians and producers is essential. Over the years, I have seen many instances of people of equal abilities applying for jobs in recording studios. Whether they followed the paths to recording engineers, or maintenance/technical engineers, has largely not been dictated by their qualifications, but by the suitability of their individual personalities. One area where this crossover of skills is of immense importance is the adjustment of monitor alignment in a room. The maintenance staff frequently know how to *adjust* the equipment, but not necessarily how to assess the sound. Conversely, the recording engineers often know the sound which they can work with, but do not know how to adjust the equipment. Co-operation is vital here in order to achieve the desired results. Obviously, many technical people do play instruments and know about music. There is, however, a distinction between playing an instrument and being a musician. That distinction once again lies largely in personality and attitude rather than in ability.

It could well be the non-adherence to this philosophy which has contributed to the low frequency inaccuracy problems of many, especially large monitor systems when used in conjunction with graphic equalisers. In the lowest two or three octaves, spectrum analysers are all but useless. The reflexions and resonances, present to some degree in *all* rooms, conspire at the lower frequencies to thwart any single-point measurement system. The use of dummy heads, and the averaging of readings from many parts of the listening area, can go some way to evening up the discrepancies, but this is time consuming, costly and still imprecise. Specialist, time-windowed systems, whilst owned by very few studios, *are* capable of measuring loudspeaker response, either with minimal disturbance by the room, or with room irregularities superimposed, but it can be nonetheless very difficult to correlate these readings with *perceived* sound character. If the engineer and producer are happy with the sound of the bass end, then go with that – irrespective of the instrumentation readings. In all too many cases, gross adjustments are made to graphic equalisers to go for a straight line on the analyser. This can be disastrous, attempting merely to compensate for a set of false analyser readings which have no bearing whatsoever on the actual, audible response. In 95 per cent of cases, I believe that monitor graphics, of which I am not an advocate anyway, are used wrongly, and sometimes seriously wrongly. When questioned, the acoustic knowledge of many maintenance personnel is almost non-existent, but they should not be blamed for that, because they are not trained acousticians. Alignment of monitors just seems to have been dumped on many of them because they are frequently the only ones who know how to adjust the appropriate controls. What ultimately matters, is what the system sounds like. It is roughly the same as entering a racing car *mechanic* in a Grand Prix, on the basis of the fact that he or she knows more about how to adjust the car than a person who is 'only' a driver. In reality the mechanic does set up the car, but the decisions on the set-up depend upon the assessments of the drivers after the performance during the practice sessions. Only by co-operation are Grand Prix won. Only by co-operation can studios perform to their full potential. In either case, a great deal of skill, experience and expertise is required. If in doubt, call in a specialist; there is so much more to many aspects of recording than meets the eye. If you would not think of calling in the acoustics specialist to repair a mixing console, then why call in an electronics engineer to fix a monitor system?

9.10 Need for an overview

In most industries of equal or greater complexity than the recording industry, a degree of true specialisation of staff members exists in conjunction with a degree of breadth of knowledge of the other aspects relating to the whole industry. This used to occur in the recording industry, when engineers were taught something about disc cutting in order that they would not record anything that could not ultimately be cut or played back from vinyl discs. In fact, the term 'recording engineer' harks back to a time when they were technically qualified people, though that, as such was not always a benefit from an artistic point of view. Anyhow, there is now a much greater occurrence of people in the recording industry having pitifully little in the way of knowledge of aspects of the recording process other than their own limited experience. This leads to unreasonable requests or demands being made upon other members of studio operating or studio building teams. It is becoming all too frequent to hear people demanding things which are totally unreasonable or impracticable given the laws of the universe and the current levels of human ability. Fighting nature will achieve nothing worthwhile; working in harmony with it can reap rich rewards.

Remember, recording is an art form, but it is also one which relies on some very profound realities of the physical universe. Perhaps that is one reason why it can be so interesting. This may, therefore, be a good time to take a look at some of the more physical aspects of how the loudspeakers perform in rooms, so in Chapter 10 we shall do that. Before, however, let us just consider a few differences which are often found between professional and less professional installations.

9.11 Life expectancy of equipment

We have already discussed the fact that professional equipment tends to be built for a longer working life than much 'promestic' equipment. However, the working practices of the studios can help to considerably extend the useful life of equipment. Certainly where mechanical devices are concerned, well-filtered ventilation systems are a great benefit. Dirt from the outside world and tobacco smoke are ruinous to many mechanical components, be they potentiometers, motors, capstans, rotating heads or equipment fans. The simple precaution of employing a well-filtered air supply is both inexpensive and highly effective. Rooms should be kept in a state of over-pressure, though, by blowing air in, as opposed to extracting it. If extractors *are* used, then they should be of less power than the air inlet fans, to ensure that the rooms are always fed with filtered air, even when doors are opened. In fact, whilst on the subject of ventilation, it is also highly useful in keeping things cool.

Many components, but in particular electrolytic capacitors, are subject to accelerated ageing when working temperatures are elevated. In general, an electrolytic capacitor working at $10\,°C$ will last roughly twice as long as one working at a steady $20\,°C$. Within the range of $10–70\,°C$, every 10 degree rise in temperature tends to halve the life of these components. Their deterioration causes an insidious loss of sound quality, which is disturbing. The need to provide good ventilation around pieces of equipment which run hot is thus of great importance, especially when many mixing consoles contain a thousand or more electrolytics. However, I do not mean, here, that if the temperature rises by $10\,°C$ for 5 minutes, then the capacitor life is suddenly halved, but rather that a capacitor working at a steady $10\,°C$ is likely to last twice as long as one working at a steady $20\,°C$. It is not simply a question of heat dosage, though, because temperature cycling is also damaging.

Most professional recording studios keep all of their equipment permanently turned on, with the exception of items such as tape recorders with continuously rotating parts, though these are now few. The heating and cooling on turn-on and turn-off cause expansions and contractions within the components, which causes fatigue and subsequent failure. Many people will have noticed noises in their studios or hi-fi equipment when first turned on, which tend to disappear over a period of minutes or hours. The simplest way to avoid this is to leave the equipment turned on, and at a stable temperature.

I was first made aware of this in 1970, when I heard that Thames Television, in England, had reported at 90 per cent reduction in breakdowns after adopting this philosophy. Obviously, though, we have two conflicting situations, here. The amount of time at elevated temperature reduces component life, but cycles of heating and cooling are also injurious to components. Giddings[1] suggests the following, which fits in well with my own experiences:

Equipment which runs hot, when idling, should be turned off when it is not to be used for three days.

Equipment which runs warm, when idling, should be turned off when it is not to be used for 1 week.

Equipment which runs *very* hot has been poorly installed, and some form of cooling should be retro fitted.

When hard-driven, convection cooled power amplifiers can get rather hot, but they all tend to run cold when idling. I tend to advise my studio clients *never* to turn them off. Valve (tube) equipment is an exception to these rules, though, because valves have typical finite lives of only around 5000 hours due to the fact that the cathode coatings run red hot, and tend to lose their electron emitting capacity over time.

Of course, to leave equipment on when unattended does need a well installed protection system, consisting of sensitive RCCBs (residual current circuit breakers) or 'earth leakage trips', as they are sometimes called. Well-installed modern equipment is rarely a fire hazard, but, for extra security, rooms containing much equipment can easily be fitted with temperature sensing fire extinguishers, which flood the rooms with a non-flammable gas should the temperature exceed a predetermined level, such as 60 °C at the ceiling. Leaving equipment running overnight is not a hazard if such precautions are taken, and when idling, will not draw excessive amounts of current. The cost of the extra electricity will usually be offset by the greatly reduced maintenance bills and the lost time whilst noise problems are sorted out or breakdowns are repaired.

Reference

1 Giddings, Philip, *Audio Systems Design and Installation*, Focal Press, Boston, USA and Oxford, UK (1995).

Chapter 10

Some basic acoustics of loudspeakers and rooms

During the discussions with the publishers whilst this book was being planned, there was some doubt as to whether this chapter should be included. It contains some simple mathematics, which can be worked out by most 12-year-olds, but nonetheless, it seems that even this can be off-putting to many potential purchasers. Anyhow, as some of the following chapters necessarily deal with some aspects of the performance of loudspeakers in rooms, unless we can establish some basic concepts, what we are discussing may be somewhat hard to appreciate.

So many of the myths and misconceptions which are heard, not only in domestic and project studios, but also in far too may professional studios, exist because so many of the basics are not sufficiently understood. It is no use demanding something from a sales representative which the laws of acoustics would preclude his or her company from making.

Loudspeakers are crude analogue electro-mechanical/electro-acoustic devices, which have been developed to a state where they can reproduce sound with realism that could hardly have ever been hoped for by their inventors in the early part of the twentieth century. However, there are only very limited circumstances in which they can perform at their best, and putting them in rooms is a great way to wreck much of what they are trying to achieve. However, rooms are where the vast majority of loudspeakers are used, so we must address the problems which they present.

By far the majority of project studios find themselves accommodated in rooms of modest size, and the control rooms, in particular, are likely to be in the order of $20\,\text{m}^2$ or less. What is more, they are unlikely to be more than $3\,\text{m}$ high, so this gives us a total probable room volume of less than $60\,\text{m}^3$. If left untreated, it is *impossible* to achieve anything approaching a uniform frequency response in such rooms, irrespective of whichever loudspeakers are used. For the purposes of some of the discussions to follow, it may well be a good idea to first look at some of the principles which govern the performance of loudspeakers in rooms. So, here we go, but in deference to the equation phobia, I will try to keep the chapter as short as possible.

10.1 Some basic acoustics

Somewhat unfortunately for acousticians, the speed of sound in air at room temperature, and hence the wavelength of any given frequencies, are more or less fixed. There are no clever 'tricks' which can change things when it would be convenient to be able to do so. The simple formula that links them is:

$$\lambda = c/f$$

where:

λ = wavelength in metres
c = speed of sound in air at 20 °C in metres per second – 344
f = frequency in hertz

Therefore:

For 100 Hz, the wavelength is given by:

$$\lambda = \frac{344}{100}$$
$$\lambda = 3.4 \text{ metres}$$

For 10 kHz:

$$\lambda = \frac{344}{10,000}$$
$$= 0.0344 \text{ m}$$
$$= 3.44 \text{ cm}$$

One wavelength is the distance taken for a whole cycle of a pressure wave, at the speed of sound, to pass through its pressure and rarefaction half cycles, as shown in Figure 10.1. The old term for hertz was cycles per second, which gave the frequency as the number of whole cycles accomplished each second. Most people reading this will already know that a 180 degree phase change gives a polarity reversal, and that if two signals of equal amplitude are mixed together, but are 180 degrees out of phase, then they will cancel. Well, the old term cycles per second comes from the cyclic nature of the phase changes, and the actual definition of frequency is the rate of change of phase, with time. A 360 degree phase rotation therefore equals one cycle. Three hundred and sixty degrees is full circle, i.e. back to 0.

If we were to put a pair of loudspeakers in a small empty, reflective room, the sound would bounce around in a manner of the utmost complexity, and what would be heard would be the direct sound from the loudspeakers, followed only a few milliseconds later by the onset of myriad reflexions, which would rapidly coalesce into what is generally referred to as reverberation. The ability to hear any detail in this confusion would be lost – the 'builders' radio' syndrome … though perhaps without the distortion. In fact, now that I come to mention it, there have been many times when builders, working on control room constructions, have been absolutely fascinated by just how good their radios began to sound as a control room progresses from its reverberant iso-

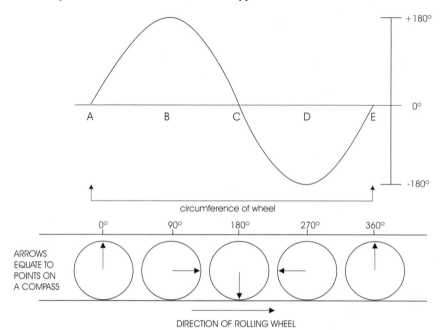

Figure 10.1 Sine wave, amplitude and phase. If the circumference of the wheel is equal to the period of the sine wave (A to E), then as the wheel rolls, a line drawn radially on the wheel will indicate the phase angle of the associated sine wave. This is why phase is sometimes denoted in radians – one radian being the phase angle passed through as the wheel advances by its own radial length on its circumference.

Therefore:

$$360° = 2\pi \text{ radians}$$
$$(\text{i.e. circumference} = 2\pi \times \text{radius})$$
$$1 \text{ radian} = \frac{360°}{2\pi}$$
$$= 57.3° \text{ approx.}$$

lation shell to its fully controlled final form. I have noticed this many times, and in many different countries. So often they begin with an attitude of 'What is this nonsense that we are being asked to do, (as acoustic structures can be rather unconventional), only to eventually admit to being both surprised and impressed by the sound of their radios in the finished room.

10.1.1 Loudspeaker directivity

In a 60 m^3 room, with moderately reflective walls of whatever shape it may take, the sound that would be heard at any point in the room away from the source would be the frequency balance of the reverberation, which would relate to the total power radiated by the loudspeakers. Due to the physics of sound generation and propagation, the frequencies below about 300 Hz tend to radiate in all directions (omni-directionally) and the frequencies above tend to radiate in a gradually narrowing beam, as shown in Figure 10.2. The figure shows lines of equal sound pressure level (SPL) which at point X all coincide.

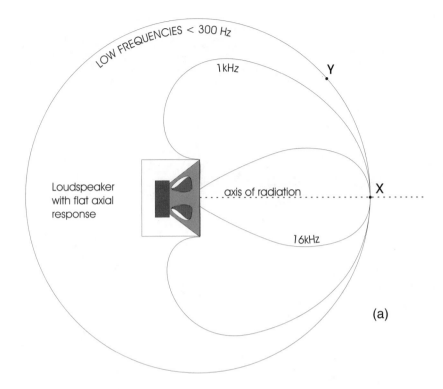

Figure 10.2 (a–d) Directivity vs frequency. (a) For a typical studio loudspeaker. Contours of equal pressure in free-field conditions. In designs for domestic use, reference point X may sometimes experience a rising high frequency response, often intentionally engineered to compensate for the greater low frequency content of the reflexions from conventional room boundaries. When designed for a flat, axial, anechoic (free-field) response, however, and fed with equal level tones, all frequencies would exhibit equal sound pressure level at point X, though at point Y a loss of higher frequencies would be experienced.

Clearly, if the distribution of sound becomes wider as the frequency lowers, then the frequencies with the widest distribution will be the ones most widely reflected under conventional listening conditions. The further relevance of this is shown in Figures 10.3 and 5.2

This would mean that a listener at point X, or at any point between point X and the loudspeaker, known as the axis of radiation, would receive all frequencies in equal balance. A listener at point Y, however, would hear the 100 Hz content of the sound at a similar level to that at point X, but the higher frequencies would be severely attenuated. The so-called frequency response of a loudspeaker usually only relates to what is heard directly in front of it, whereas the reverberation and reflexions in the room are driven by all the sounds which radiate in all directions, most of which do not have flat responses. It is worth noting here, that this is most definitely *not* what happens with acoustic instruments.

If a listener at point X were to perceive an even quantity of the three frequencies shown, then there would need to be an equal source intensity travelling between the loudspeaker and the listener along the path between the

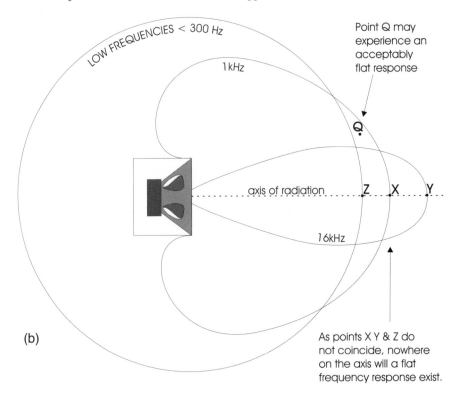

LOW FREQUENCIES < 300 Hz

1 kHz

16kHz

axis of radiation   Z X Y

Point Q may
experience an
acceptably
flat response

Q

(b)

As points X Y & Z do
not coincide, nowhere
on the axis will a flat
frequency response exist.

Figure 10.2 *cont.* (b) For a typical domestic loudspeaker. Contours of equal pressure.

In order to take into account the extra low-frequency room response, many domestic loud-speakers are designed to have more high frequencies. Reference point X may often experience a rising response, intentionally engineered to compensate for the greater low frequency content of the reflexions from conventional room boundaries. However, by not having a flat axial response, such loudspeakers *cannot* have an accurate transient response, because the transient signals will arrive, with their rising frequency response, before any room reflexions can return to restore the overall tonal balance, yielding a very 'bright' sound. In large rooms, where the delay before the arrival of the first reflexions is greater, the effect will be more pronounced. In smaller rooms with little acoustic treatment, however, such an option may be more subjectively acceptable. In a relatively dead room, the general response (not only that of the transients) of a loudspeaker such as this would be unacceptably bright, because it would be trying to compensate for the effect of room reflexions which did not exist

two. This would imply a uniform (flat) axial response – the axis being the imaginary line between a listener and a loudspeaker when the listener is directly facing the radiating surface of the loudspeaker – the dashed line in Figure 10.2. If equal power must radiate along the axis for a flat axial response, then it should appear obvious from the above figure that at 1 kHz, the loudspeaker needs to radiate more power than at 16 kHz, because the contour of equal pressure covers a greater area. It follows that at 100 Hz, even more power is radiated, because the contour of equal pressure at 100 Hz covers a greater area still. If one spreads the same thickness of butter over a larger slice of bread, then one needs to have more butter in order to do so, or spread it thinner.

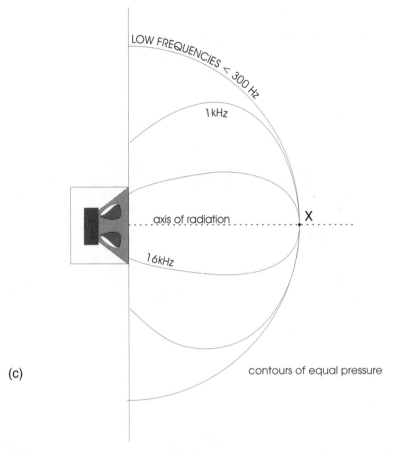

(c)

LOW FREQUENCIES < 300 Hz

1 kHz

axis of radiation   X

16kHz

contours of equal pressure

Figure 10.2 *cont.* (c) For a flush mounted loudspeaker. By comparison to (a) and (b), it can be seen that there is no rear radiation. The low frequency energy is forced forwards, and therefore must be reduced in level if point X is not to experience a bass boost. The useful outcome of this is that around 3 dB less power is required, which increases the loudspeaker headroom. This means that it can either go 3 dB louder than when free-standing (at low frequencies) or, for the same SPL, the low frequency loudspeaker will be operating at only half the power, and thus will be suffering less strain. The loudspeaker shown in (a), if flush mounted, would exhibit a response as shown in (d) if no measures were taken to equalise the response

Figure 10.3(a) shows the response of a loudspeaker on its axis, and at 15 degree intervals off axis, up to 60 degree, and Figure 10.3(b) shows the total power response for the same loudspeaker. Note the gradually rising power as the frequency lowers, in order to maintain a relatively constant pressure on axis. The axial and off-axis responses were measured in an anechoic chamber, which is virtually free of any reflexions. The chamber used is shown in Figure 10.4, and is rated down to 70 Hz, below which the absorption becomes less than sufficient to call the room anechoic. The total power response of Figure 10.3(b) was made in the reverberation chamber shown in Figure 10.5. To avoid the response being affected by the standing wave pattern in the room,

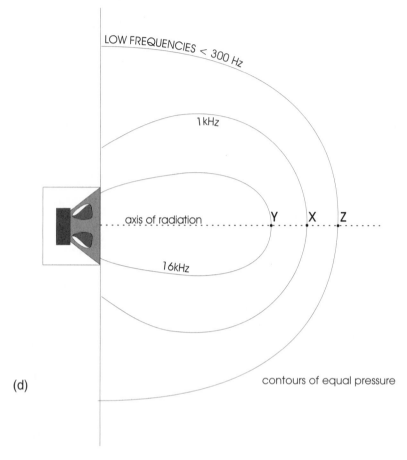

(d)

Figure 10.2 *cont.* (d) Note here that points X, Y and Z are in the opposite order to the similar points in (b). This is why many manufacturers of loudspeakers for professional use, with in-built amplifiers and electronic crossovers, provide level and tilt controls for the low and high frequencies. This allows adjustment for achieving a flat axial response (or whatever other response may be desired) in a wide range of environments.

It should now be obvious from (a) to (d) that *no* loudspeaker can be designed to have desired response unless its mounting conditions and the room acoustics are known in advance

the response was summed from measurements made with the loudspeaker in two different positions. During each measurement, the microphone was moved around the room, in a continuous and random motion, to try to achieve some spacial averaging. This particular chamber has an average reverberation time of about 8 seconds.

As the reverberation chamber tends to sum the total output (the movement of the loudspeaker and microphone was only to smooth over any minor irreg-ularities), clearly it would not be possible to measure axial and specific off-axis responses in such a room. The plots would all tend to look like the total power plot. Conversely, in the anechoic chamber, it is not possible to measure total power at any one point. This could only be done by measuring at hun-

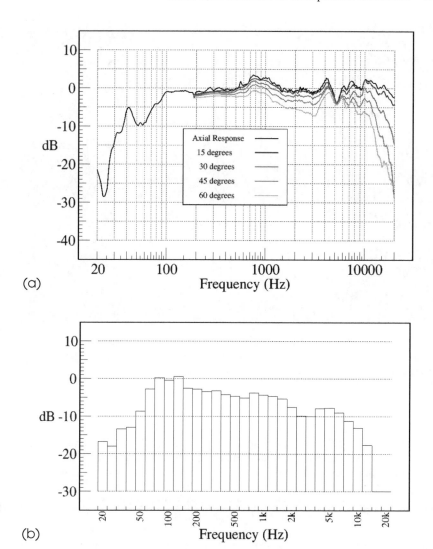

(a)

(b)

Figure 10.3 Different response aspects of one loudspeaker. On-axis, in an anechoic chamber, the perceived response would be that of the axial response, shown in (a). However, as the off-axis plots show, the perceived response at other listening angles is likely to sound short of top. In a highly reverberant or reflective room, the perceived response will be more like the total power response shown in (b), which shows the frequency balance that would be heard at almost all places in the room. The total power response plot, as shown here, was actually made from a measurement in a reverberation chamber

dreds of points at a fixed distance from the source, and integrating all the measurements; a very tedious exercise indeed. The reverberation chamber reflects the radiation in many (more or less all) directions back towards the microphone with only minimal losses, so it does the hard work for you. In the

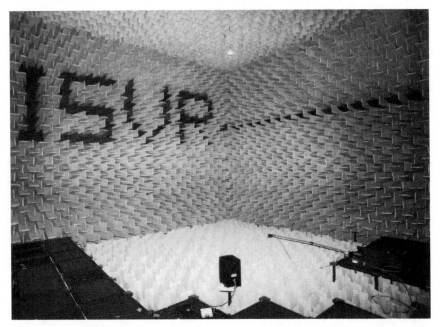

Figure 10.4 ISVR anechoic chamber. The large anechoic chamber in the Institute of Sound and Vibration Research at Southampton University, UK, has a volume of 611 m³. The chamber is anechoic down to around 70 Hz. Below this frequency, the wedges represent less than a quarter of a wavelength, and absorption rapidly falls off. The floor grids are completely removable, but are strong enough to support the weight of motor vehicles

chamber shown in Figure 10.5, the mean distance travelled for any sound before it strikes a surface is about 6 metres, so at 334 metres per second for 8 seconds, each sound would reflect about 450 times before decaying by 60 dB.

$$\text{Number of reflexions} = \frac{\text{speed of sound (m/s)} \times \text{time (s)}}{\text{mean free path (m)}} = \frac{334 \times 8}{6}$$

The reverberation time, $RT_{60,}$ is the time taken for a sound to decay to 60 dB below its original level. This was a practical level when these standards were set, because an orchestra would peak at around 100 dB in a concert hall, and with a typical background noise level of around 40 dB, the decay of the big finale would be lost in the noise once it had fallen by about 60 dB. Sixty decibels also represents a fall to one-millionth of the original power.

10.2 Realistic listening rooms

In practice, rooms of any normal nature tend to fall somewhere in between the properties of the anechoic and reverberation chambers. They therefore exhibit some of the properties of each. When listening to loudspeakers in a normal room, if one listens very closely in front of the loudspeakers, one

Figure 10.5 The large reverberation chamber at the ISVR. It has a volume of 348 m³ and a reverberation time of 10 seconds at 250 Hz, and 5 seconds at 2500 Hz. The pipe, visible in the photograph, was in use supplying high pressure air to a horn which is capable of producing 180 dB and shattering metal panels. All surfaces are non-parallel, and are made from painted concrete

tends to hear the axial response as the dominant characteristic. As one moves away from the loudspeaker, and further into the room, the response will tend towards the total power (reverberant) response. The distance where the response changes from being predominantly axial, to being predominantly reverberant is known as the *critical distance*. For rooms of identical structural shapes, sizes and materials, the critical distance will differ according to the amount of absorption in the room. In practice, this absorption is achieved by soft furnishing, such as carpets, curtains, sofas and chairs, and at the lower frequencies, also by windows, doors, floorboards, plaster ceilings, book-cases and virtually anything else which may be in the room which is not both solid and massive. The energy used to vibrate the objects is absorbed before it can become reflected energy.

10.2.1 Infinite variety

The point which should now be becoming clear is that unless any two rooms are identically constructed and identically furnished, and contain people of identical shapes and sizes, then they will each exhibit different degrees of absorption, and so will exhibit different critical distances, which will also be frequency dependent due to the frequency dependent nature of the different absorbing surfaces. It thus follows that listening to identical loudspeakers in different rooms, especially at any distance beyond the critical distance, will yield different perceived frequency balances. Even within the critical distance, responses would vary because the critical distance is defined as that distance beyond which the reverberant response *dominates*, not as the distance beyond which the reverberant response is just audible. In a room with a strong response, the room resonances may be audible even with one's nose pressed against the loudspeaker.

What is more, the position of the loudspeakers and listeners in a room will also affect how the room is driven; and how its response is received. This is due to the coupling to the resonant room modes and the reflexion paths. This adds more variability to the equation. Worse still, the reflexions which return to the loudspeaker cones create differing pressure fields against which the cones must push. When a cone has more to push against, it will do more work, and thus will radiate more energy. Moving a loudspeaker within a room will change the relationship of the reflexions returning to the cone(s) so will change the radiated response – the power response. This will be discussed further in Chapter 12, as it has much impact on multiple loudspeaker systems and panned images.

With all of the above variables playing such important roles in the overall response of a pair of loudspeakers in a room, it is little wonder that there should be such a degree of room-to-room, and system-to-system variability, and that so many people's home hi-fi systems sound so different. In fact, it is one of the reasons why there is such a range of loudspeakers available on the hi-fi market. In many cases, the wide-range of loudspeakers, in any given performance range, is the only way that an overall desired response can be achieved from the huge quantity of different listening environments.

10.2.2 A common reference

The degree of variability in domestic rooms is something that must be lived with. It is a reality of life. Nevertheless by mixing in a more restricted range of room variability, it is possible for people, in their homes, to optimise their choice of loudspeakers and their positions in the rooms in order to obtain a most pleasing or accurate response. However, this is only possible if the majority of recordings are mixed in rooms which share many common characteristics. Otherwise, when optimising a home hi-fi for one set of recordings, all it may serve to do is to show up the corrections which were made to tonal balances and relative instrument balances to compensate for the anomalies of poor mixing room acoustics in the studios.

For example, let us presume that one studio has a dip in its response at 100 Hz, and another has a peak at the same frequency. Let us now also suppose that some fortunate person has found a set of conditions at home which have yielded a reasonably flat response. The personnel mixing in the room with the 100 Hz dip would tend to boost that frequency band in order to compensate for their lack of ability to hear it in the control room. When played in the more uniform domestic acoustics, the subsequent recording would sound to have a peak at these frequencies which were boosted. The music mixed in the control room with the *peak* around 100 Hz would tend to be mixed with those frequencies reduced, and so would sound lacking in that region when played on our aforementioned domestic system. Any profession which puts out such

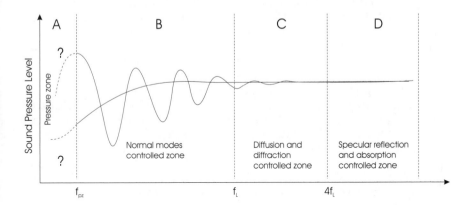

Figure 10.6 The diagram shows the frequency ranges over which different aspects of room acoustics are the predominant controllers. Explanations of f_{pz} and f_L are given in the text. In the pressure zone, the response is not supported by room effects. Despite its uniformity, the loud-speakers must do all the work, here, so the loudspeaker driving pressure zone frequencies will need to be capable of considerable low frequency output. The gently curving line represents the average response of the room (after Bolt, Beranek and Newman)

inconsistent results is not doing justice either to itself, or to its customers. Professional control rooms, which include project studio control rooms which are there to make commercially available recordings, should, at least, do whatever they reasonably can to ensure that their monitoring conditions are relatively uniform. Achieving uniformity, though, in a small room, is some-what more tricky than in larger rooms.

10.3 Room control

Figure 10.6 shows the different frequency regions which are governed by the different acoustic properties of rooms. When a sound wave strikes an object there are three basic things which can happen to it: it can be transmitted, it can be reflected, or it can be absorbed. The same things occur with light when striking a window. The light which enters a window is transmitted through the glass. Standing outside the window, we will be able to see ourselves in the glass; this is, of course, a reflexion – light which is sent back in the direction of its origin. Inside the room, if we then open the window, we will find that a little more light gets in. The difference between the light which passes through the glass and the light which enters when the window is opened, *minus* the light which is reflected back to its source, represents the light which is being absorbed *within* the glass, and turned into heat.

Sound can also be diffused and diffracted. Diffusion has another light analogy. If we put frosted glass in our windows, then light energy will enter, but no detail of the source will be apparent, and only vague shadows will be cast inside the room. Diffusion intermingles the discrete energy sources, and spreads them widely. Diffraction is the bending of the sound waves as they pass around objects, particularly things with sharp corners. Again, diffraction occurs with light, which diffracts around the edges of an opaque body. The diffraction of light waves, as with sound, can be frequency dependent, and can produce rainbow effects as light passes through a narrow slit, or around a sharp corner. Of course, there is nothing strange about there being such close parallels between light and sound, because they both deal with wave motion. In general, the same laws of wave motion apply to the electromagnetic waves of light or radio, the acoustic waves of sound, or the mechanical waves of the ocean. Wave motion is wave motion.

Anyhow, back to Figure 10.6. At high frequencies, the sound level in a room is a combination of the direct sound, and a balance between the sound which is specularly reflected by hard objects and that which is diffused and diffracted. Specular reflexions are ones which return to the room both discrete and intact. A light beam, pointed at a mirror, would give a specular reflexion. A light beam shone at a white sheet, such as photographers use to reflect light on to a subject, would produce a diffuse reflexion. This is what takes place acoustically in the second higher frequency band, labelled C in Figure 10.6. Diffraction effects (the wave bending around objects) also have a great influence in this frequency band. Lower down the scale, in section B, the characteristics of a room tend to be controlled by the room modes, where resonant energy can augment the direct energy, sometimes producing unpleasant, resonant hang-overs. These are often referred to as standing waves. 'Resonant modes' is a better term, though, because standing waves of a totally benign nature can also exist.[1] In all the above areas, the direct sound is augmented by the various effects; the diffracted, diffused and reflected energy adding to the 'loudness' of the direct sound. In the modally controlled region, the perceived frequency response will tend to follow the peaks and dips, hence producing the irregularities on certain bass notes, as discussed a few paragraphs earlier, where different rooms exhibit boosts and dips at 100 Hz.

The very lowest frequency region of Figure 10.6 is the pressure zone. This region exists below a frequency whose half-wavelength is greater than the longest dimension of the room. The formula for finding this is very simple:

$$f_{pz} = c/2Lr$$

where:
f_{pz} = pressure zone upper limit
c = speed of sound, in metres per second
Lr = longest room dimension, in metres

Thus, for our typical $5 \, m \times 4 \times 3 \, m$ room, the upper limit to the pressure zone would be:

$$f_{pz} = \frac{344}{2 \times 5}$$
$$= 34.4 \, Hz$$

In this region, we could be assured of a very flat frequency response. The total room/loudspeaker response would be that of the loudspeaker, as determined by the loading imparted upon it by the room. Remember, rooms are pressure vessels, just like loudspeaker cabinets, so just as a cabinet loads the rear of a loudspeaker, so does a room load the front. The pressure zone is therefore not supported by any room effects, so despite the uniformity of its response, pressure zone frequencies will tend to be lower in level in most rooms, compared to those in the region above, which are supported by reflected energy.

10.3.1 Large rooms

The first room modes tend to be well-separated in frequency, so an untreated room would tend to have a response more like the wavy line in Figure 10.6, rather than a more correctable response as shown by the 'average room response' line on the same plot. By making the room large, the separated modes can be pushed down to lower frequencies, around 17 Hz for a 10 m room, and the pressure zone upper limit will also be reduced in frequency. By damping the modes, by means of absorption, the roller-coaster effect of the response can be reduced. Some degree of damping is required in any control room, otherwise not only would the pressure variation be both frequency dependent and position dependent, but at resonant frequencies, the energy would 'hang on' in the room. In such circumstances it would be difficult to know whether the actual sound of a bass drum decayed slowly when being monitored, or whether the slow decay was down to the resonances of the room in which the monitoring was taking place.

The positional effect is demonstrated by Figure 10.7, which shows the pressure distribution in a room being driven by a 70 Hz signal. The darker areas are of higher sound pressure variations from the norm. Either a loudspeaker and/or a listener sited in the darker areas would generate or receive very strong sounds, as opposed to when situated in the lighter areas, when those sounds contain frequencies close to the resonance; in this case 70 Hz. When a room exhibits a single troublesome resonance, there can be some flexibility in response normalisation by moving either the listening or driving positions, or both. However, because the wavelengths of the different frequencies are also different when a room has more than one troublesome resonance, the patterns of higher and lower pressure variation will probably not positionally coincide. Moving out of a trouble zone at one frequency may well just move into another one at another frequency.

The modally controlled region extends from the upper limit of the pressure zone to a frequency around what is annotated as f_L in Figure 10.6, and which is known as the 'large room frequency'. Again, this can be calculated by a very simple equation:

$$f_L = K\sqrt{RT_{60}/V}$$

where:

K = an SI constant (here 2000)
V = room volume, in cubic metres
RT_{60} = room decay time to –60 dB, in seconds

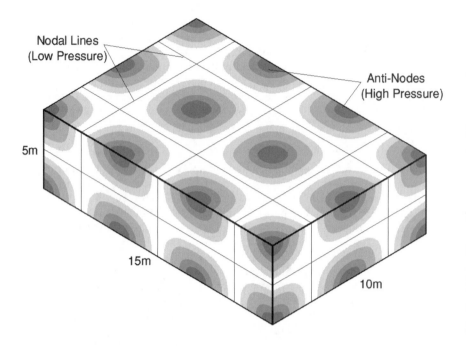

Figure 10.7 Distribution of a 70 Hz resonance in a room. The dark areas show the positions of maximum pressure variation (anti-nodes) and the white areas depict the regions of minimal pressure variations (nodes). The more reflective the room surfaces, the more will be the contrast from light to dark. An anechoic chamber would show a rather uniform greyness, when similarly excited

For a 5 m × 4 m × 3 m room with an RT_{60} of 1 second, the upper limit of modally controlled region would be given by:

$$f_L = 2000 \sqrt{\frac{1}{5 \times 4 \times 3}}$$
$$= 2000 \sqrt{\frac{1}{60}}$$
$$= 2000 \sqrt{0.0166}$$
$$= 2000 \times 0.129$$
$$= 258 \, \text{Hz}$$

In our 5 m × 4 × 3 m room, therefore, we have a modally controlled region between 34 Hz and 258 Hz, or thereabouts, so these would be the frequencies in most need of absorption if we were to try to flatten the response. In larger rooms, where there is plenty of space for absorption systems, these problems can easily be addressed to a significant degree. Also, in larger rooms, the reflective energy will have further to travel on its way from the source to the listener, so it will be more separated in time and will be lower in level when it returns to the listener. Large rooms also have more surface area available for absorption, which is another factor in their favour.

Just as an example, a room of 10 m × 8 m × 5 m would have a volume of 400 m^3, a pressure zone frequency of 33 Hz, and a large room frequency (f_L) given an RT_{60} of say 1.2 seconds of:

$$f_L = 2000 \sqrt{\frac{1.2}{400}}$$
$$= 2000 \times \sqrt{0.003}$$
$$= 2000 \times 0.0547$$
$$= 109.5\,\text{Hz}$$

In small rooms, the modal band thus intrudes to a much greater degree.

In certain types of larger control room designs, temporal and spacial separation are combined with diffusive and absorptive treatments to provide some very pleasing and admirably smooth acoustics, but as wavelengths and arrival times are all fixed by the speed of sound, these concepts, which rely on some critical psycho-acoustic functions, often will not scale into smaller room dimensions. This was mentioned in the opening paragraph of Section 10.1.

10.4 Summary

What this chapter should by now have shown, albeit in only a fraction of its true complexity, is how loudspeakers radiate sound in a very non-uniform way (totally different, incidentally, than the way in which acoustic instruments radiate sound) and how rooms respond to the sounds, also in very non-uniform ways. The following chapter will now seek to demonstrate a very effective way of dealing with these problems, and how to achieve high degrees of monitoring uniformity and clarity, even in rooms as small as 14 m^2 and 36 m^3.

Reference

1 Newell, P.R., *Recording Spaces*, Focal Press, Oxford (1998)

Bibliography

Beranek, L., *Acoustics*, McGraw-Hill, London (1974)

Borwick, J., *Loudspeaker and Headphone Handbook*, 2nd edn, Focal Press, Oxford, UK (1994)

Colloms, M., *High Performance Loudspeakers*, 5th edn, John Wiley and Sons, Chichster, UK (1997)

Newell, P.R., *Recording Spaces*, Focal Press, Oxford, UK (1998)

Chapter 11

The small room problem

In the upper echelon of the world's sound recording studios, the current trend is towards large control rooms. This has been prompted by the evolution of the recording industry, which has not only led to larger mixing consoles and more equipment, but also, since the mid-1970s, to the 'invasion' of the control rooms by musicians. Furthermore, the subsequent development of midi synchronisation gave rise to the tendency not to record many of the keyboard instruments, but only to programme them during the recording process, then to feed them directly into the mixing console during the mixdown process. So, with many current mixes involving mixing consoles of a hundred channels, or more, a virtual mountain of effects processor, and stacks of keyboards, the shift towards larger control rooms is not surprising.

Large control rooms have other benefits, though. Acoustic control is more readily achieved, partly because the resonant room modes (or 'standing waves' as they are frequently referred to) tend not to become isolated from one another until much lower frequencies than in small rooms, and also because the reflexions which return from room boundaries do so both later in time and lower in level. The reflexions which have travelled further are more attenuated when they arrive at the ear. Their later arrival also allows the ear to recognise them as separate events, and therefore they are not so readily confused with the direct signals. This is not to say that good large rooms are easy to design, they are not, but they do begin with fewer limitations than small rooms, and, additionally, large rooms are usually built with budgets that can afford serious design. In small rooms, however, life can be much more difficult. The boundaries are in close proximity to the ears, and the mountains of equipment often grow vertically, where they will more readily interfere with the monitoring. What is more, the physical volume of the equipment is more of a dominant factor in smaller rooms, giving the acoustics less space to 'breathe'. A large mixing desk in a small room drastically affects the uniformity of the response of the monitors, even in rooms which have been well designed and which have excellent responses when only lightly loaded.

Nevertheless, small control rooms have been a fact of life since the earliest days of electrical recording, and today they are probably the norm in the most rapidly expanding portion of the recording industry, the project studios. Economic considerations and the problems of finding suitable studio premises in convenient locations will almost certainly ensure that small control rooms continue to be built. What is more, and *very* importantly, there is an increasing amount of work in multi-media facilities, which almost always have small control rooms for sound mixing, and all too often suffer (put up with?) less than optimum sound monitoring conditions. Unfortunately, sound is still usually seen as the poor relation in the multi-media family, but there are finally signs that better sound is now being called for. All of this now demands a radical re-appraisal of what can be done about providing better, more consistent, and more trustworthy monitoring conditions in small control rooms, because the problem of compromised monitoring is likely to increase unless the small room problem is seriously and rapidly addressed.

11.1 Room size and modal behaviour

By 'small' I am referring to rooms of less than about $100 \, \text{m}^3$, and rooms of that size would typically be around 6 m × 5 m × 3 m high. Remember, small, in acoustic terms, is frequency (wavelength) dependent, so a room which may be considered large where a response only down to 50 Hz is required, may become small if the response needs to be extended down to 20 Hz. A room becomes acoustically small when the energy in the resonant modes, which are functions of the size and shape of the room, fails to overlap. Modes, for those people unfamiliar with their concept, can be thought of as the pathways between reflective surfaces which the sound waves travel. When the sound can bounce back and forth, returning in phase to its starting point, the energy can build up, and the resonant 'standing wave' is formed. For resonance to occur, the distance between any surfaces must correspond to whole wavelengths, and hence specific frequencies relate to specific path lengths. Once modal separation occurs, different frequency components of any music reproduced in the room will be heard at different levels, dependent upon whether or not they coincide with, and receive support from, the natural modes of the room. When the modes are overlapping, the frequency response will be more or less uniform, but once separation occurs, it will become irregular. The overall response of the room begins to follow the shape of the individual modes. (See Figure 11.1.) The separation begins at higher frequencies in smaller rooms, it so follows that larger rooms will exhibit a more uniform response down to lower frequencies, all other characteristics of the rooms being equal, that is.

The other major drawback about resonant room modes is that the room response becomes position sensitive, both for the source, such as a loudspeaker, and the point of reception of the sound, such as the listener(s). The directivity patterns of the source will also influence which modes are driven, and which are not. When either the source or listener is at a point of minimum pressure variation (a node) of any given mode, then for the frequencies which coincide with that mode, either nothing will be transmitted, or nothing will be received. Only in truly anechoic spaces will a room allow the unmodified response from the loudspeaker to arrive at the listener without different distances between them producing different perceived low frequency responses. What complicates life even more is that each room reflects energy back into itself, not only according to its dimensions, but also according to the materials and methods of construction. Different structures absorb different frequencies

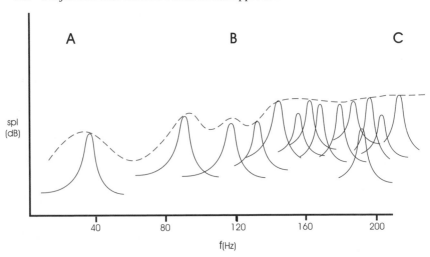

Figure 11.1 Modal separation. The plot represents the energy contained in individual modal resonances in a room. In the region between B and C, the modes are reasonably closely and evenly positioned. The 'best fit' average level is shown by the dotted line, and is relatively uniform. However, in the region A–B, the resonant modes of the room are well separated in frequency, and the dotted line, which represents the 'frequency response' of the room, now tends to follow the individual peaks and troughs of the modal energy. It can thus be seen that as the modes separate, the room response will become irregular, favouring some notes more than others. The frequency of point B, where the modes begin to separate, is a function of the room dimension, lowering as the size increases. Large rooms therefore tend to have a more uniform response down to lower frequencies

to greater or lesser degrees. The absorption, and associated acoustic damping, will determine the strength with which any reflected energy will return to the room, and will also affect the Q, or energy spread, of the resonant modes (see Figure 11.2). The Q, in this sense, is like the Q of an equaliser, where the peak can be either broad (low Q) or narrow (high Q). Q, incidentally, stands for 'quality factor', but in our case, a poor quality resonance (low Q) is desirable, as it is less well defined, so is usually less audibly intrusive than a high Q resonance. Highly damped, absorbent rooms hence both spread the frequency content of the modal energy, and reduce the peak levels. All of this suggests, and it is in fact the case, that the perceived response in an absorbent, low Q, room, would be more uniform than in a room with harder walls and modes of higher Q, though the latter would sound louder for any given acoustic input because it would be slower to dissipate the energy. The practical extreme of the absorbent, highly damped rooms, are the anechoic chambers, which, because of their excellent response uniformity, are used for measurement purposes.

Unfortunately, at low frequencies, absorption becomes difficult in small rooms, in which the size of many low frequency absorption systems would preclude their use. Effective low frequency absorbers have traditionally been large; their dimensions often requiring a depth of one-quarter of the wavelength of the lowest frequency which must be absorbed. As a 40 Hz wavelength is around 8 metres, that would mean using an absorber system 2 metres

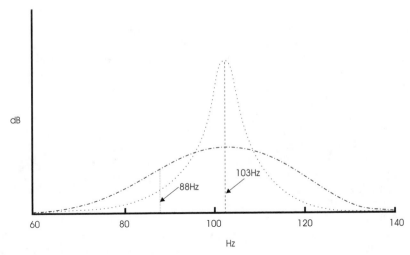

Figure 11.2 The effect of damping on 'Q'. The dotted line represents a high Q resonance which will be strongly excited by stimulus at 103 Hz but will respond only weakly to a stimulus at 88 Hz. The dashed line shows the effect of damping in lowering the Q. In this case, excitation at either 88 Hz *or* 103 Hz will produce an almost identical degree of resonance. Both the dashed and the dotted curves represent resonances with a 103 Hz centre frequency, and both contain, for clarity, a similar amount of total energy. In reality, though, due to absorption, the damped mode would contain less energy, so would be lower in level. It can thus be seen that the low Q resonance responds to a broader range of frequencies than the high Q resonance, but it will not respond as strongly to any common excitation.

deep. Where the bass absorption in a control room is not adequate, then the perceived level of different low frequencies will be dependent upon the listening and source positions, and if there remains an excess of low frequency, semi-reverberant (reflected) energy, then the room will sound 'bass heavy'. If the room is a recording room, then this fact may, or may not, be desirable, dependent upon the instrumental and musical requirements. In a control room, though, a bass-heavy room will tend to lead to bass-light mixes, which may sound weak when played elsewhere (See Section 10.2.2). In another bass-heavy room, a bass-light mix will probably sound more like what was intended, but 'bass heaviness' can also be frequency dependent, and some musical keys may be perceived as being more bass-heavy than others, with different rooms favouring different keys. This is clearly undesirable in a control room, where response uniformity is a fundamental requirement.

11.2 Current concepts and bad influences

Resolving the problem of room-to-room compatibility is not easy, and as the rooms get smaller, the problem tends to increase. When the reflected/'reverberant'/diffusive energy is perceived only from the back of the room, as is the case with a number of different techniques of control room construction, then if insufficient time is allowed between the first pass and the reverberant and/or reflected energy returning to the listening position, the psycho-acoustics upon which the rooms rely may not be effective in producing the desired results, and a coloured response will be perceived. Some of these designs, such as the 'Live End, Dead End' room, are quite complex in nature, and can perform well when carefully designed, but there have been too many cases where people have tried to copy the concepts and scale down the size of such rooms, without fully understanding the principles involved, and these practices have led to some really awful small rooms.

Whilst the widest and most uniform 'frequency' response would seem to be the common aim of all recording studio control rooms, this worthy and entirely logical goal would not appear to have achieved even remotely universal acceptance. The biggest problems lie in the middle-order studios, which, although they produce the bulk of the world's recordings, exist in a highly competitive marketplace where expensive and/or space consuming solutions to their acoustical problems are often out of the question. They also often exist in a market where peer pressures and practices, rather than absolutes, dominate their thinking. I have known studios who wanted improved monitoring conditions, but when these have resulted in a sound which was unlike the most commercially successful studio in the area, they have been faced with the problem of their clients becoming unsure about their new situation. It is likely that the other studios were successful, *despite* their monitoring, due to a host of other appealing aspects, such as friendly and attentive staff, good studio rooms, easy parking facilities, and so forth. However, the recording industry is full of insecure people, which is a relatively normal situation when people are working on a creative edge, and especially where success or failure can rest entirely on subjective assessment. Surely though, one would think that when such conditions prevail, islands of security, such studios with repeatable and consistent monitoring, would be jumped at, but the only believable security often seems to be following somebody else's perceived success, howsoever gained, which can, in itself, be very volatile.

There is also an increasing tendency to think that problems are solved by electronic equipment. For example, the poor sound on a recording is frequently not perceived to be due to poor decision making (perhaps as a result of inadequate monitoring), but because the studio did not possess the latest valve pre-amplifier, or a new electronic reverberation device. The great majority of the larger studios know better, of course, which is why they invest in acoustics. Many of the owners of the smaller studios seem to hope that future success will bring new, bigger premises, in which they would consider some acoustic control to be obligatory, but they are often very reluctant to spend money on acoustic treatment for their small studios. The fact that it could significantly increase client satisfaction, business in general, and help to bring closer the day when a new premises *could* be afforded, is often offset by the fear of not being able to recover the money of the work and materials invested when the move was made, but this attitude takes its toll on quality. Despite all this, the reality still exists that good, repeatable, 'accurate', detailed monitoring cannot be achieved in poor rooms. Where good monitoring does exist in a poor room, by fluke, it will almost certainly be improved in a better room.

11.3 A different outlook

So, how can we deal with these complex problems in a way that is not only capable of producing much more controlled monitoring conditions, but will also not be disproportionately expensive, either in relation to the overall cost of equipment, or to the hourly charges which the studio can make. Well, the first thing we should do is to look at the three main variables in our small room response: source position, listening position and the nature of the room itself. If each and every source location, in anything other than an anechoic chamber, is going to drive the room modes differently, then it would seem that the best thing to do is to try to fix the loudspeaker positions at points of minimum variability. Even a large anechoic chamber, unless it had absorbent wedges of 3 metres or more in length, would be unlikely to be anechoic down to the lower octaves of the audio frequency range, so even here, the different in-room locations would produce different bass responses. The most uniform way of driving a room is from its boundaries, and loudspeakers mounted in the front wall of a room meet this criterion. Within the standing wave field of a room, the boundaries form surfaces of maximum pressure (anti-nodes), so a loudspeaker set in to a boundary will drive all such modes equally, and will hence not favour the excitation of some modes more than others. Flush mounting of the loudspeakers also allows the wall to act as an extension to their baffles, and this is useful in producing more uniformly expanding sound waves. The reason *why* the boundaries of the rooms are surfaces of maximum pressure is because in order for the sound to reflect from them, it must first change direction. This involves coming to a momentary stop, and when the particle velocity is zero, the pressure must be at a maximum, or physical laws relating to the conservation of energy would be violated.

As discussed in Chapter 10, almost all loudspeakers become more omni-directional as the frequency lowers, so in the case of free-standing loud-speaker cabinets, the lowering frequencies will be able to radiate all around the boxes, and will, if the loudspeakers are situated *within* the room, find their way to the wall behind them. They will subsequently be reflected back into the room, and to the listening position (See Figure 5.2). The different frequencies will have different wavelengths, and hence, given that they have a common path length from the loudspeaker to the wall and back to the listener, they will return with different phase relationships. These will tend to either add to, or cancel, the direct response, and will lead to an unevenly perceived response at the listening position. It is true that the wall behind the loud-speakers could be made absorbent, but at low frequencies this would both consume much space and reduce the amount of radiated output from the loud-speakers arriving at the listener. Also, if *all* of the room surfaces were to be made absorbent, then we would end up with an anechoic room, and such rooms are not pleasant to be in. Many people even find them disturbing.

If, however, the loudspeakers are mounted flush in the front wall, and if this wall is solid and non-resonant, it will aid the creation of a uniform pressure distribution in the room, and it will also increase the radiated low frequency output of the loudspeakers by forcing all the pressure forward. The effect can easily be seen by comparing Figures 10.2(a) and 10.2(d). This can be advantageous in extending the low frequency response, especially when only limited headroom is available in the monitor system. If it is also made to be acoustically reflective, it will provide a source of 'life' for the speech and actions of people within the room. It *cannot*, however, produce unwanted monitoring reflexions, as all the sound from the loudspeakers is radiating *away* from the wall. If any loudspeakers which were built-in were designed for free space use, there would be a relative increase in the axial low frequency response, but this could be corrected electrically because it is a minimum phase effect; amplitude correction would not upset the phase response – indeed, it would correct it. (See glossary for minimum phase.) Remember from Chapter 5, the difference between the on-axis boost created by the restriction of the radiating angle and the boost caused by the arrival of any more or less in-phase reflexions, is that the reflexions suffer an arrival delay due to the greater distances which they have travelled. Such boosts are neither uniform with frequency, nor are they arriving at the listener at the same time as the direct sound. These effects, therefore, cannot be compensated for by electrical equalisation, and it is the ill-informed attempts to use such correction that has given a bad name to the use of equalisation on monitors.[1,2] Electrical equalisation, other than of the digitally adaptive type, cannot fix room mode problems, and in all cases, making one place in the room better will make another place worse. Reflexion problems are acoustic problems, which need acoustic solutions. It would thus seem to be a minimum (mandatory?) requirement for 'accurate' monitoring, where a quality control function is required, for the loudspeakers to be flush mounted in the front wall, in *any* control room. Chapter 12 will deal with some of these electro-acoustic phenomena in more detail.

If the loudspeakers cannot be mounted in the structural wall, which may in any case be bad from the sound isolation point of view, then a dense false wall should be provided. A flimsy wall would be absorbent, and would resonate and re-radiate at certain frequencies, so this would only add another disturbance to the response. Consistent and uniform radiation of the sound is the first step on our way to good monitoring, but this will not achieve its goals unless their is a uniform energy distribution in the reflective or the semi-reverberant sound fields. In larger rooms, different designers can have different techniques (and can follow different philosophies) in their approaches to ensuring a perceptually uniform response at the mixing console. In small rooms, though, because reflexions can return from different locations with different frequency balances *and* at relatively higher levels than in larger rooms, due to the close proximity of the walls, these reflexions can return within the psycho-acoustic integration time of the brain. They can thus be perceived not as confusing reflexions, which are bad enough, but as timbral coloration of the sound. The situation is made worse by the non-uniform off-axis frequency response of the loudspeakers, which creates further frequency imbalances in the reflected sound.

On the other hand, if *all* the surfaces of the room, with the exception of the front wall, and perhaps the floor, are made absorbent, then effectively it is only the direct signal from the loudspeakers which is perceived by the listener(s). Fortunately, it is perhaps the easiest of the tasks facing loudspeaker designers to get a uniform axial (±30°) response, so loudspeakers designed for such rooms can often be realised more economically than loudspeakers which aim to be usable in the widest possible range of rooms. Much loudspeaker development, and hence cost, has been consumed by the attempts to produce systems which can exhibit relatively flat frequency responses in poor rooms, but in fact, in monitoring circumstances, such wide directivity monitors are only a palliative. The published responses for such loudspeakers will almost certainly have been made using anechoic, or reflexion-free, measuring techniques, so this, in itself, suggests that even the manufacturers accept that the responses will be worse in other situations. The aim of what is being proposed in this chapter is a situation where the anechoic *and* listening position responses will be essentially similar. The basic concept of such rooms is shown in Figure 11.3.

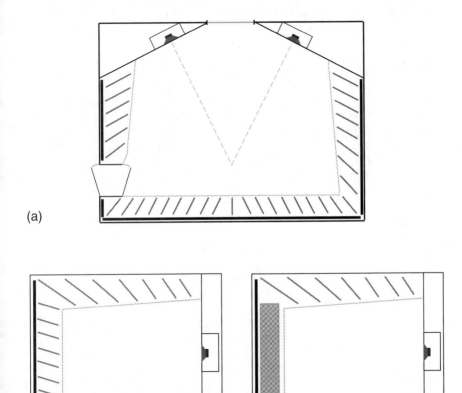

Figure 11.3 (a) and (b) show the essential features of the Non-Environment room concept, in which only a single pass of the sound waves emitted by the loudspeakers is permitted to travel past the ears. A hard front wall is used in order to provide 'life' to the speech and actions of people in the room, to relieve any oppressive anechoic experiences. Nevertheless, the rooms approximate to anechoic terminations for the loudspeakers, for which the front wall acts only as a baffle extension. (a) is a plan of an NE control room, with the shaded areas denoting wide band absorber systems; (b) shows the side elevation, with shading denoting (1) horizontal rear absorbers and (2) vertical rear absorbers

11.4 Furnishing disturbances

Of course, any equipment placed within the room will also produce reflexions, but not too much imagination is needed to site the equipment such that it will not reflect the sound directly to the listening position. However, perhaps a little more should be said here about reflexions from equipment, as their behaviour is frequency dependent. Objects which have dimensions that are small, compared to any given wavelength, will tend to be engulfed by the sound, and it will pass around them, 'swallowing' them. On the other hand, when the surfaces of objects are large in comparison to the wavelength, they will tend to act like mirrors. Sound travels at about 340 metres per second in air at around 20 °C, so a 340 Hz sound wave will have a wavelength of about 1 metre. A 100 Hz wave will hence have a wavelength of over 3 metres, and such a frequency will have little trouble in 'swallowing' a small, desk-top loudspeaker with a maximum dimension of 40 cm. In the case of a 10 kHz sound wave, though, with a wavelength of only about 3 cm, the sides of the loudspeaker box will be many wavelengths wide, so this will reflect, somewhat like a light-reflecting mirror. Thought of another way, if a large, bouncy ball is thrown against an irregular surfaced wall, then if the irregularities are small in comparison to the dimensions of the ball, the ball's surface will tend to engulf the irregularities, and it will bounce back in a direction similar to that in which it would bounce from a flat wall. If a small ball is then thrown against the same wall, and if the irregularities are of equal or larger dimensions than the ball, the ball will tend to bounce off, at an angle which is dependent upon the angles of the faces of the irregularities which it strikes. This is not an exact analogy, but it does give something of a feel for the wavelength dependence of reflexions. Low frequencies engulf mixing consoles, but when consoles have large, flat backs, consideration should be given to how the lower mid-frequencies will respond to them, and any potential for chatter between the front wall and the back surface of the mixing console should be dealt with by absorbant material on the back of the console.

11.5 Origins of the concept

The concepts of the rooms being described here were a development from Tom Hidley's 'Non-Environment' rooms, which he began building in the mid-1980s, and continues to do so to this day. Many of his control rooms are huge, but this is because he is choosing to push his responses down to 10 Hz, and at these wavelengths, absorption in small spaces is impossible. Tom and I co-sponsored a post-graduate research project into low frequency absorption, in 1990, at the Institute of Sound and Vibration Research at Southampton University, and, due to the fact that many of the concepts of acoustics will scale, the work aimed at improved absorption at very low frequencies led to the possibilities of absorbing more 'normal' frequencies in smaller rooms. The concepts of the philosophy of these rooms were presented in a paper given to a conference of the UK Institute of Acoustics in 1994[3].

The philosophy behind these rooms aims to monitor, as closely as possible, what is on the recording medium. Very briefly, if the range of possible listening conditions is so great, from headphones to discotheques and motor cars, then to what reference should a control room acoustic comply? The internationality of the music market now largely makes the concept of an average listening room redundant, as the acoustics of typical homes vary enormously from country to country. The recorded signal now seems to be the only relevant thing to monitor. Furthermore, many different types of music are better suited to different listening environments, but no control room can be all things to all people. The 'Non-Environment' approach transfers the subject of listening acoustics to the final listening environment (room, car, headphones etc.), as in reality, compromises between these environments usually lead to fudges. Carefully designed large rooms *can* produce some well-chosen compromise performances, but once again, small rooms are a different matter. Of course, it will be clear from this paragraph that I am *not* an advocate of the 'mixing to the market' philosophy. I believe we should mix to a standard.

The absorbent walls and ceilings of these rooms are multi-layered, absorbing the impact of the sound waves gradually, so not to create reflexions from the sudden transition from the air to the absorbent. Just like electrical reflexions from mis-matched terminating impedances, acoustic absorbents will reflect energy if there is a sudden change in the acoustic impedance. This is one reason why anechoic chambers use wedges, so there is a gradual change from air to foam, or glass fibre, or whatever the wedges may be made from. It is also similar to the modern concept of battle armour. Fifty centimetres of steel plate has now given way to lighter weight, multi-layered, and more effective protection materials, which do not produce such an abrupt termination to the flight of the incoming projectile.

11.6 Structural principles

Figure 11.4 shows the first layer of the rear wall absorption system of a small control room. In the finished item, this is usually covered by an acoustically transparent fabric, but the first layer (from the entering sound wave direction) consists of suspended sheets of chipboard, covered on one side with an acoustic damping layer, and on both sides with 5 cm of a fibrous absorbent material. Between the panels, there are air spaces. These are complex impedance devices, which absorb in many different ways. Behind these panels, there is usually a large panel, of similar construction, hanging parallel to the back frame of the acoustic control shell. The shell would typically consist of stud partitioning frame with 5 cm to 10 cm of fibrous material inside the frame. The side of the frame facing the loudspeakers would be covered with an acoustic deadsheet and another fibrous layer. The side of the frame which faces the structural walls of the room is usually covered with a sandwich of plasterboard and acoustic deadsheet. The deadsheet, on the inside of the sandwich, forms a 'constrained layer' which greatly increases the acoustic damping. This causes acoustic energy loss, and helps to spread broadly the frequency content of whatever modes remain in such a control room. Finally, the space between the acoustic shell and the structural wall is partially filled with an absorbent fibrous material, such as Rockwool, Paroc, or Noisetec A1 from Acoustica Integral SA. Almost any fibrous material of about 30–40 kg/m^3 would be effective, though. In the case shown in Figure 11.3, the total wall depth (thickness) is around 60 cm, but a full scheme is shown in Figure

Figure 11.4 Rear wall absorber system

11.5. The arrangement of a typical ceiling absorber system is shown in Figure 11.6. (Readers unfamiliar with deadsheets should refer to the glossary.)

The principle of operation is to allow the sound waves to enter the 'traps' relatively easily, then to 'nibble away' at the energy by different processes. The traps act partially as lined ducts, partially as panel absorbers, and partially as membrane absorbers. The mechanisms of loss are many. The large panels act as typical panel absorbers, and being made of various composite materials, create frictional losses within the material of their construction. There are also losses due to tortuosity, both on a macro scale, and on a micro scale. On the macro scale, the waveguide effect of the angling of the front panels exposes the sound waves to a greater surface area of absorbent than would be the case for a direct impact, and, by causing the waves to strike the absorbent at an oblique angle, causes them also to pass through a greater depth of the fibrous material, which under most circumstances increases the absorption. The deadsheet covering of the chipboard increases its damping, and also helps to prevent any re-radiation which would occur if the panel were to continue to vibrate for any significant period of time after the impact of a transient. The return path to the room through the labyrinth of panels is somewhat convoluted, and in itself creates even more loss.

On a more micro scale, the losses within the fibrous material are due partially to tortuosity losses, as the acoustic wave must find a much more complex, and less direct path through the absorbent than it would through air. There are adiabatic losses which are caused by a 'heat sink' effect which robs the air of the heat of compression and the cooling of rarefaction, both of which normally contribute to the speed of propagation of the sound wave through the

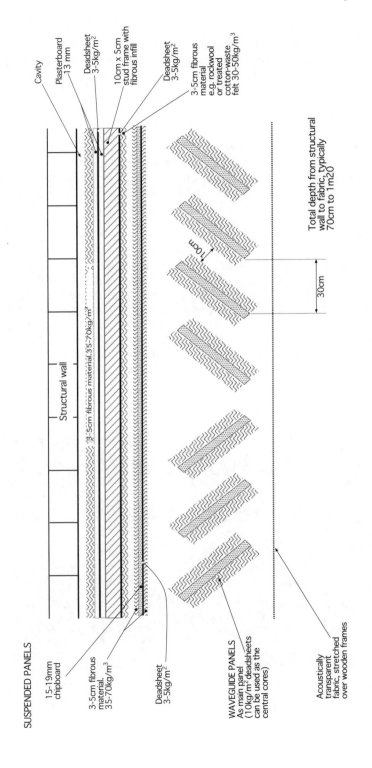

Figure 11.5 A typical rear wall absorber system

air. Viscous losses arise from the 'stickiness' of the air, as it rapidly tries to change direction, both within the fibrous material, and in the spaces between the panels, which create the lined ducts. The panels themselves hang freely, so that any energy which enters the panels largely remains to dissipate within them, and is not transmitted into any other parts of the shell from which it could be re-radiated. Coupling to any structural modes is thus avoided.

The large panel absorbers work mainly on the low frequencies, and these panels are further damped by the proximity of their rear fibrous covering to the deadsheets, which form the inner lining of the shell frame. If these panels are pulled forward, then left to fall, the air damping can clearly be seen as the panels almost stop dead when close to the vertical position, and take some time to squeeze out the last of the air from the space. The deadsheet lining of the walls forms a limp bag, which absorbs the low frequency blows. The damping is further improved by the trapped air in the cavity of the frame. The outer sandwich of plasterboard and deadsheet provides another low frequency absorbing layer, and is heavily damped. What passes through all of this must then cross the cavity to the structural wall. The fibrous lining of this cavity reduces the tendency for any resonant modes to form in the space, particularly the ones parallel to the wall surfaces. What is reflected from the outer, structural wall will then have to pass through the whole 'trap' system again, in the other direction, before it can return to the listening position. This effectively doubles the effect of the trap system (though *not* its effective depth), and severely attenuates any modal activity in the room. Incidentally, the word 'trap' was first used for the forerunners of these systems by Bart Locanthi, at

Figure 11.6 Ceiling absorber panels

JBL, in the 1950s, and for some time it was considered to be a rather loose description. However, during research being carried out at the Institute of Sound and Vibration Research, in 1998, even some members of staff were beginning to consider that it is perhaps rather an appropriate word, after all.

The behaviour of the structural wall is entirely dependent on its construction, but certainly some sound will pass through it, some will be absorbed within it, and some will reflect from it. Anything absorbed in the structural wall is of benefit to both sound isolation and internal acoustic control. On the other hand, transmission through the wall is good for internal control, but bad from an isolation point of view. The reverse is the case for its reflective properties. This is one reason why 'stock' plans for studios cannot compete with individually designed rooms, as the effect of the structure can play a significant part in the design considerations. Each studio also has its own set of circumstances regarding noisy or noise sensitive neighbours.

11.7 Philosophical irrationalities

These rooms are neither boxes full of Rockwool, nor the end of acoustic design (as one critic referred to them in the international recording press in 1996) but are capable of providing very effective solutions to some difficult acoustic problems. In fact work continues into finding new techniques for wide band absorption in less space. As the problem being discussed in this chapter largely concerns *small* rooms, then by definition, there is little enough space to begin with, so any treatment cannot be too space consuming. What seems strange, though, is that so many studio owners seem to be reluctant to lose *any* space in exchange for a much better sound. Many of them want to *see* in the finished result as many as possible of the square metres that they are renting or buying. It is a puzzle, to me, how they fail to appreciate that they are *hearing* the benefits of the whole space; and that the fact that they are *sound* control rooms evidently seems to count for little. It is quite frightening how many studio owners, given the same 6 m × 5 m floor area to start with, opt to use a relatively untreated 6 m × 5 m room that sounds like nonsense, rather than a 5 m × 4 m room, after treatment, which sounds exceptionally well controlled. I have no idea where this mentality springs from, or how these people manage to be called professionals, but the industry is loaded with them.

Perhaps it is the 'It's all in the equipment' philosophy again. In many cases their clients have been led (by many glossy advertisements) to believe that no studio should be without certain pieces of equipment, so they demand that the studios should have the latest devices. Unfortunately, the clients', musicians' and producers' knowledge of acoustics is often minimal, certainly in the middle order of things, and they are often relatively unaware of what they may be missing. True, clients often complain about the monitoring in many studios, but it seems that on the great majority of occasions, they suggest different loudspeakers, and not better rooms. Usually, in these cases, they have either heard some of the preferred loudspeakers in an entirely different room, or have a CD which they think is great, and which because they saw a photo in a magazine of the studio in which it was recorded, presume that the loudspeakers which they see must be 'the ones'. This may sound somewhat cynical, but it is not meant to be so. Unfortunately, it is based on some all-too-common reality.

11.8 Conclusion

Basically, the problem of the absolute consistency of small room monitoring conditions is insoluble in terms of conventional acoustic control. The acoustic can only be removed to as great a degree as possible. The moving coil loudspeaker is now over 70 years old, and the fact that the problems of interfacing it to the modal distribution of rooms is still taxing the minds of many great acousticians suggest that the problem is not trivial. However, the principles being described in this chapter can very effectively produce small control rooms having response uniformities which are commendably close to those of the axial, half-space radiation response of their monitor loudspeakers. This method can produce excellent room-to-room compatibility, and room-to-outside-world predictability, but it is, of course, being presumed here that control rooms are generally required to have uniform responses, and that they will be required to perform some 'quality control' monitoring task. I realise that there are idiosyncratic creative environments for recording, but at the stages of mixing, and certainly when mastering, a more uniform approach is required. If greater consistency is not achieved, then the current situation of engineers circulating around many studios will have little chance of giving consistent productions to the record buying public. The newer trends into multi-media formats will only serve to complicate the situation further, unless the small room problem is rigorously addressed. Fortunately, the control room techniques described here lend themselves well to the control of rooms which are somewhat overloaded with equipment. In multi-media studios, where, in acoustic terms, an excess of reflective surfaces already exist courtesy of the piles of equipment, the hard front wall can be replaced by a more absorbent surface. This can help to control the 'chatter' between the wall any hard surfaces at eye level, which tend to be more usual in multi-media rooms than in purely sound control rooms.

The 'Non-Environment' style of room can function from rooms of 40 m³, to rooms of 2000 m³, and still present remarkably uniform monitoring. They do require good acoustic design in the most appropriate application of the absorber systems for any given room, *and* in the design of the absorber systems themselves. However, these room designs are *not* very sensitive in terms of the sizes, shapes and angles which must be so very carefully calculated in the case of many other room design concepts. The fact is, these rooms work; as music control rooms, as film dubbing theatres, as television control rooms, as mastering rooms, and for most other applications where a quality control requirement exists. They have steadily been building up an enthusiastic following world-wide; and for good reason!

References

1 Newell, P. R., 'Monitor Equalisation and Measurement', *Studio Sound*, Vol. 34, No. 9, pp. 41–51 (September 1992)

2 Newell, P. R., *Studio Monitoring Design*, Focal Press, Oxford (1995)

3 Newell, P. R., Holland, K.R. and Hidley, T., 'Control Room Reverberation is Unwanted Noise', *Proceedings of the Institute of Acoustics, Reproduced Sound 10* Conference, Vol. 16, Part 4, pp. 365–373 (1994) (Republished in 2 above)

Chapter 12

Stereo, the unstable illusion

Perhaps an apology is in order to any faint-hearted readers before embarking on this chapter, but just because life can be complicated is no reason to hide from it. Usually, it is far better to confront one's problems. It may be that I am labouring the subject of loudspeakers in a book that purports to cover a wide range of recording topics, but it has been my experience that if there is any, one thing that preoccupies the interests of people involved in recording equipment, then it is loudspeakers. Remember also, that the loudspeakers are the window through which all recordings are ultimately viewed and judged. Forgive me, then, if I launch into a difficult chapter, which will make rather unpleasant reading for anybody seeking a simple life.

Despite loudspeakers being the subject of so much interest, there are many aspects of their performance which are taken for granted, yet are nonetheless totally misunderstood. In later chapters, we are going to discuss some aspects of surround sound, but it is difficult to get some of the points across if stereo is not fully understood. In Chapter 10 we looked at some of the acoustic aspects of loudspeakers and rooms, and in Chapter 11 we considered some of the difficulties and how to overcome them. Unless we get some fundamentals understood, if only partially, then we may be wasting some good opportunities for getting some information to where it can be useful. I have been 'reminded' several times to bear in mind the title of the book, and to be aware of the levels of understanding of much of the likely readership; but talking down to people seems to be a great way of insulting their intelligence and keeping them in the dark.

I would suggest, therefore, that if any reader gets a little overwhelmed by this chapter, they should persist with it. If, by the end, they have only muttered to themselves a couple of times 'Well, I never realised that', then it will have achieved something – at least. If any readers get arrested by 'men in white coats' for throwing their beloved stereo systems out of their windows, then we will have achieved even more. If the result is an urge to read on, then we will have broken new ground, and we may be on our way to creating some new engineers, who have a genuine right to call themselves such.

In the March 1996 issue of *Studio Sound*, John Watkinson wrote a thought provoking article about the concepts of the specifications for a perfect loudspeaker. The perfect loudspeaker is, indeed, a laudable aim, but in reality, even if it could be realised, it might not turn out to be quite as useful as many people would perhaps expect. A perfect loudspeaker is one thing, but its application, especially in stereo pairs, is something else entirely. In the same edition of *Studio Sound*, Tomlinson Holman, the 'TH' of 'THX', wrote an article on multi-channel stereo, and touched on a subject which creates endless problems for loudspeaker designers and studio designers alike. The subject referred to was acoustic coupling, not only between loudspeakers and rooms, but also between the different loudspeakers in a multi-channel system. Tom Holman suggested the rule 'pan first, then equalise', and there is a great deal of sense in what he says, as under most common circumstances, the 'one side only' response will differ from that of a phantom centre image. Unfortunately, even when a theoretically perfect loudspeaker *pair* encounters the unavoidable effects of acoustic coupling, we begin to run into problems. It is some of those problems which we will discuss in this chapter, because, once understood, the concepts of the 'ultimate' or 'perfect' stereo monitoring system are dealt a fatal blow.

12.1 Mutual coupling of loudspeakers

A stereo pair of loudspeakers should be considered as just that; a stereo pair, and *not* two mono sources. As we shall discuss later, there are situations where the pairs refuse to unite, but first, perhaps we should look at a mono pair before extending the discussion into the realms of stereo. Let us consider the mounting, side by side, of two 15 inch loudspeakers in one cabinet. Such an arrangement will produce substantially more low frequency output than a single, identical driver, in a similar cabinet, receiving the same electrical input power. Of course, the pair will handle twice the input power; and the lower impedance, if they are connected in parallel, will usually allow the amplifier to deliver more power to the pair, but these are not the subjects under discussion here. We are only considering here the additional low frequency output produced by supplying 100 watts into the pair, compared to 100 watts into the single driver. The mutual coupling between the two drivers will produce more radiated acoustic power from the pair than would be produced by the single driver receiving the same total input power.

If we consider the output from one loudspeaker only, its response will be affected by the air loading, which is a result of the way in which the air moves out of the way as the diaphragm moves. When we fit two loudspeakers side by side, each one has the effect of locally varying the air pressure on the other. Whereas one loudspeaker, on its own, is loaded only by the air which is constrained by the walls of the room in which it is mounted (or by free air if outside) two drivers working together are each loaded by the locally constrained air, *plus* the extra pressure variations from the other, in-phase connected driver.

If we think of two, equally rated boxers, one swinging punches into a heavy bag, and the other one swinging punches into a bag of loose feathers, the boxer hitting the heavy bag will start sweating much sooner than the one hitting the feathers. The reason is that the heavy bag provides more resistance to the blows, gives the boxer more to push on, and so causes more work to be done in moving the bag. When a loudspeaker moves forward into the air, the air provides some resistance to its movement, and against this resistance the work is done which converts the motion of the radiating surface into acoustic energy. If this resistance is increased, such as by adding pressure from an adjacent radiating surface, more work will be done, and more sound will be radiated.

Similarly, horn loading a loudspeaker gives it a greater conversion efficiency by making it less easy for the air to get out of the way when the radiating surface (cone, diaphragm, or whatever else) moves. The increased output, on axis, is totally the result of the extra load provided by the air constrained within the horn. In the case of two non-horn loaded loudspeakers in one box, the additional output is due to the superimposition of the radiated pressure from one driver increasing the load upon the other. This is why a 2 × 15 inch monitor loudspeaker will have more LF output than a 1 × 15 inch version, even when receiving the same total input power.

Let us now examine the performance of a stereo pair of direct radiating monitor loudspeakers mounted in the front wall of a moderately reflective/reverberant control room, with their centres spaced 3 metres apart. Let us also consider these loudspeakers to be of an ideal type, which, under the given loading conditions, each produce a perfectly flat frequency response from 10 Hz to 40 kHz. Bearing in mind the aforementioned 'pan first' rule, we will listen to, and measure, the response of a bass guitar played through the left loudspeaker only. If we then pan the signal fully right, we will hear, and measure, the same results as the 'left only' test. However, should we then pan the image into the central position, there will be heard to be a relative power increase in the low frequencies, plus, perhaps, some coloration in the upper bass region. The LF boost can be up to 3 dB (see Figure 12.1), but will vary from one room to another depending on the nature of each room's reflectivity. It will also vary according to the exact nature of the drive units, and the cabinet loading on the drive units.

A 3 metre separation equates to a wavelength of around 110 Hz, so in the case of a stereo pair of 15 inch loudspeakers separated by 3 metres, all frequencies below the *half* wavelength frequency, about 55 Hz in this case, will be boosted. This boost can be due to both the extra loading on each radiating

(a)

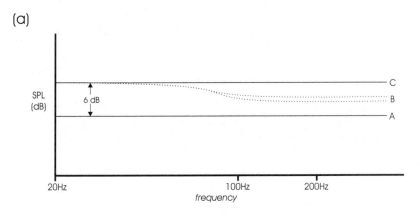

Figure 12.1 (a) Pressure amplitude responses in an anechoic room. A: Response of a single loudspeaker anywhere in the room. B: Response of a stereo pair of loudspeakers, each receiving the same input as A, anywhere in the room, except on the central plane (precise response may be position dependent). C: As in B, but measured on the central plane

(b)

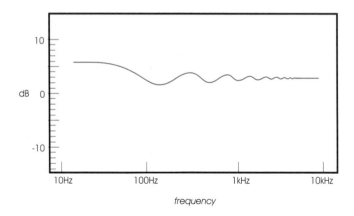

frequency

Figure 12.1 *cont.* (b) Frequency response of the same pair of loudspeakers as in (a) at any position in a reverberant chamber (combined power output)

(c)

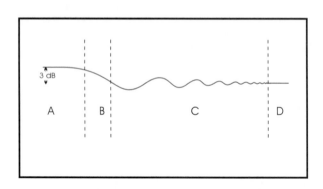

Figure 12.1 *cont.* (c) Zones of loading. General response as in (b). Zone A: region where the
separation distance between the loudspeakers is less than half-wavelength distance, and where
the two loudspeakers are close to each other's near fields. Zone B: region where the separation
distance between the loudspeakers is less than the half-wavelength distance, but where the
outputs are beginning to reduce with distance as the frequency rises. Zone C: region where the
separation distance is greater than the half-wavelength distance, and where the mutual coupling
alternates, with frequency between being constructive or destructive. Zone D: region where the
mutual coupling has ceased

surface, produced by the pressure from the other loudspeaker in the pair
(hence *mutual* coupling), and also from the reflexions arriving back at the
loudspeakers, from the room boundaries. The situation that we have here is
that below the half-wavelength separation frequency of 55 Hz, a centrally
panned image would effectively be radiating from a 2 × 15 inch monitor
system, but our left or right panned signals would be emanating only from

1 × 15 inch monitor systems. The mutual coupling would therefore tend towards giving the central image a 3 dB boost at low frequencies. The boost will, in practice, be somewhat less than the 'perfect' 3 dB, as beyond the acoustic near-field of a loudspeaker, the pressure falls off by 6 dB for each doubling of distance, but reflected energy can also augment the response at very low frequencies. At frequencies around 50 Hz, however, two 15 inch loudspeakers in one cabinet, such as in many large monitor systems, would be well within each other's acoustic near-field, so would effectively act as one, single source, with the possibility of the full 3 dB of increased output, *plus* whatever the room reflexion loading may add.

It should be pointed out, here, that the mutual coupling effect can occur at all frequencies, and is either constructive or destructive dependent upon whether the signals arrive at the opposite loudspeakers in phase, or out of phase. This is a function of the wavelength and the distance between the radiating surfaces, but it is independent of the listening position. It is not like the cancellations and summations caused by arrival time differences in the listening area, which *are* position dependent. However, when drivers are sited less than one half wavelength apart, and if they are connected in phase, the mutual coupling between them becomes entirely constructive; though reflective coupling from the room boundaries may arrive at the left and right loudspeakers with different phase relationships.

The mutual coupling can only take place when each of the loudspeakers is located in a position such that the sound pressure from the other can impinge directly on its radiating surface. In anechoic conditions, the effect is highly dependent upon the loudspeaker positions and their directivity patterns. At frequencies below those where the distance between the drivers is less than half a wavelength, the coupling would be entirely constructive. The extra power radiation would be proportionately less for frequencies where the SPL reduces with distance. Above the wholly constructive mutual coupling frequency (the half-wave separation) the summation would be either constructive or destructive, or anything in between, dependent on wavelength and distance, and hence the phase at the point of arrival at the opposite loudspeaker. The implication here is that *only* on the centre line between two loudspeakers can a stereo phantom image exhibit a flat frequency response. Remember this, it is very important. I am going to say it again. A centrally panned image can only exhibit a flat frequency response on the centre line between the loudspeakers. The proof of this is shown in Figure 12.6.

12.2 Reflective coupling

There is a time-honoured way of considering an acoustically reflective room in the same way as an optically reflective one. Figure 12.2 shows a mirrored room, in which a listener would hear a reflected sound from each place where an image of the loudspeaker was visible. This happens in all three dimensions. We therefore have a situation whereby, in conventional rooms, the real loudspeakers also acoustically couple to the 'phantom' reflected loudspeakers. So, to re-cap, we have the direct mutual coupling affecting our response, and we also have the disturbances in the low frequency loading which are caused by the boundaries of the room constraining the angle of radiation.

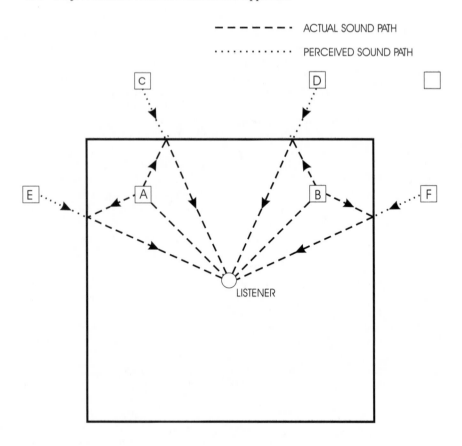

Figure 12.2 Mirrored room analogy. Reflexions behave as if they were independent sound sources, located at the positions of their images. Floors and ceilings behave similarly. **A** and **B**, actual sound sources; **C–F**, apparent sound sources

These are both loading effects on the radiating surfaces, and are independent of listening position. Furthermore, we have the disturbances of the far-field response caused by reflexions from the room boundaries. The latter are of a non-minimum phase nature (see Glossary, if necessary) as there are delays inherent in their arrival at the listening positions, in which case they cannot be compensated for by conventional equalisation. Their effects are also highly dependent on both the source, *and* listening positions.

The theoretically ideal loudspeaker is a very valid starting point, and for the purposes of this discussion it is a useful concept, but it should now be coming clear that perfect loudspeakers would fail to *perform* perfectly in conventional rooms. The complex room interactions, which were previously more prevalent in the control rooms of the 1960s than they are today (in well designed rooms, at least, but in many project studios, perhaps they still remain) tend to produce an overall, on-axis response which is more akin to the 3 dB, total power summing, rather than the 6 dB summation which is usually expected on-axis in modern control rooms.

12.3 The pan-pot dilemma

The perception of a 3 dB, overall central summation was at the root of the old pan-pot dilemma: should the electrical central positions of a pan-pot produce signals which are 3 dB or 6 dB down, relative to the fully left or right positions? Mono electrical compatibility of the stereo balance requires constant voltage, therefore each side should be 6 dB down (half voltage) in the centre, in order to sum back to the original voltage when added *electrically*. (The pan-pots, or more fully, *pan*orama *pot*entiometers are *pot*ential (voltage) dividers, so, it is the voltages which must sum.) On the other hand, in the case of an *acoustic* stereo central image in a reasonably reverberant room, it is the *power* from the two loudspeakers which must sum to unity in the centre, and which therefore requires a condition whereby the output from each loudspeaker would be only 3 dB down (half power) when producing a central image. Nevertheless, below the mutual coupling frequency, Figure 12.1(b) shows how the low frequency response would still approximate to the –6 dB requirement, as the power summation would tend to be augmented by a further 3 dB due to the mutual coupling effect between the two loudspeakers.

In stereo radio drama, where voices are often panned across the sound stage, –3 dB pan-pots would (in conventional rooms) produce a uniform level as the voice was panned from left, through centre, to right. However, if the programme was then to be broadcast in mono, the voice would be perceived to rise by 3 dB as it passed through the centre position in the stereo mix. If the same task were to be repeated for a bass guitar, then in the mono broadcast it would still be subject to the same, uniform, 3 dB rise as it passed through the centre of the stereo mix. Somewhat inconveniently, though, when heard in stereo, it would be perceived to increase only in its low frequency content as it passed through the centre position on its way from left to right. Hence the importance of the advise, 'pan first, equalise later'.

As there is usually little dynamic panning of low frequency instruments in stereo music recording, a –3 dB centre position used to be considered optimum for stereo music recording. In the case of radio drama, where mono compatibility for the majority of the broadcast listeners is a great priority, a –6 dB centre position would be required. Many mixing console manufacturers will produce consoles with different pan-pot laws for different applications, though they usually opt for a –4½ dB compromise, which produces only a 1½ dB worst case error in either instance. The fact that this seems to work well is borne out by the number of recording engineers who fail to realise that this situation exists at all. Anyway, in real life, movements tend to lead us to *expect* some variability in level, so the psychology of human perception works in our favour here.

What is more, most listening rooms, and especially control rooms, now tend to be rather more 'dead' than 'live', so they approximate more to the electrical summing conditions. In fact, in truly anechoic conditions, the acoustical sum on the central plane is identical to the electrical sum, at least in the region where the axial response holds true; but, this is a central plane-only condition. Elsewhere in the room, the summation approximates to the panning effect at all points in a more reverberant space, though with less confusion in the sound. This is another aspect which has led to the development of control room philosophies which require very dead monitoring acoustics.

12.4 Room decoupling

The undoubted complexity of the interaction of the coupling mechanisms so far discussed is one of the reasons why I have strongly supported the type of control room design which was outlined in Chapter 11, the principles of which are deeply rooted in the solutions to the room to room variability of these coupling problems[1,2,3]. Using the optical analogy once again, the 'Non-Environment' rooms have mirrored front walls and floors; with all the other surfaces being matt black. With the loudspeakers being thought of as spotlights mounted in the front wall, their only visible reflexions at the listening position would be the single reflexions from the floor. This one remaining reflexion from each loudspeaker can effectively be 'smudged' by the use of

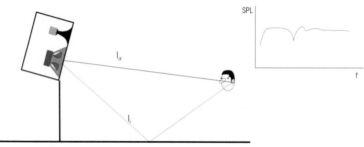

Low frequency response of a conventional 2 x 15″ loudspeaker system with woofers side by side in a typical monitor environment.

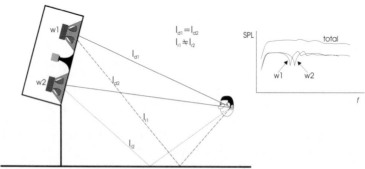

Low frequency response of a loudspeaker system with two vertically mounted woofers (bass drivers) in a typical monitor environment.

Figure 12.3 (a)Low frequency response of a conventional 2×15 inch loudspeaker system with woofers side by side in a typical monitor environment. (b) Low frequency response of a loudspeaker system with two vertically mounted woofers (bass drivers) in a typical monitor environment (after Shozo Kinoshita)

monitor systems with twin bass drivers, mounted vertically. Figure 12.3 shows the effect of such a mounting system, and shows how the resulting floor reflexion can be rendered relatively innocuous. The continuing work in the development of these designs is now in the areas of how to make the 'matt black' even blacker, especially at the lowest frequencies, and how to reduce the physical size of the absorber systems.

As the loudspeakers in such rooms are built into the front wall, their position is fixed. The mutual coupling of the left and right loudspeakers will therefore be entirely predictable. Even in very small control rooms (15 m^2) the absorber systems can now achieve great effectiveness, which leaves the loudspeaker-to-loudspeaker mutual coupling as the only significant coupling to be considered. (Although, see Appendix on p. 194.) Room to room consistency and predictability become much easier to achieve with this type of room, large or small; but there are other aspects of arrival times and double source artefacts which no room design can address: they are inherent weaknesses in the concept of two-channel stereo itself, so perhaps we should now look at some of these points.

12.5 Transient vs steady state performance of a phantom image

Perhaps it should be noted here that sound power and electrical power are equivalents, but the electrical equivalent of sound pressure is voltage. That is why we relate to 3 dB being a doubling of power, but doublings of voltage or pressure represent 6 dB rises. On the central plane of a stereo pair of loudspeakers, the transient *pressures* will sum, producing a single pulse of sound 6 dB higher than that emitted by each loudspeaker individually. At all other places in the room, as the different distances to the two loudspeakers create arrival time differences, double pulses will result. This effect is clearly shown in Figure 12.4. Although it may seem to an observer on the central plane that four times the power (+6 dB) of a single loudspeaker is being radiated, the effect of the constructive and destructive superimposition of the signals around the room as a whole would still only result in an average increase of 3 dB. However, this still apparently leaves us with a 'magic' extra 3 dB of power on the centre line, which cannot be described, as previously, in terms of radiation impedance. The superimposition of the pressure from one loudspeaker on the other cannot be the cause, because the transients fly away from their respective sources *before* the effect from either source could superimpose itself on the diaphragm of the other source. So, we need to look at this behaviour.

In the case of a perfect delta function (a unidirectional impulse of infinitesimal duration) the points of superimposition would only lie on a two-dimensional, central plane, of infinitesimal thickness. As this would occupy no perceivable space, then no spacial averaging of the power response would be relevant. This would thus not violate our 3 dB total power increase for two identical sources. With a transient *musical* signal, however, which has finite length, there would be positive-going and negative-going portions of its waveform. At the places along the central plane where the transients crossed, they would not meet at a point, but would 'smear' as they interfered with each other over a central area, either side of the central plane. Around this central plane, the pressures would superimpose, producing a pressure increase of 6 dB over a finite region each side of the centre line. As they crossed further, they would produce regions of cancellation, which would show total power losses exactly equal to the power gain in the central region of summation. The total power would thus remain constant when area-averaged. The above effect is shown in Figure 12.5, in which the average height of all the transients occurring at any one time in the room would be the same as that of a single transient, emitted by one loudspeaker, though they would be doubled in number as there are two sources. Only where they interfered with one another would there be disturbances in their height, but there would be no extra total power in the room, just the simple sum of the power radiated by the two individual loudspeakers.

On transient signals, therefore, such as drumbeats, because of their exis-

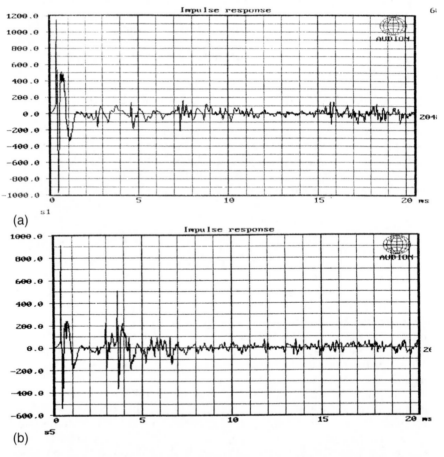

Figure 12.4 (a) Impulse response from a centrally panned image from a pair of loudspeakers, measures taken on the common axis of both loudspeakers (centre line). The response from a central, mono loudspeaker would be essentially similar. (b) Impulse responses as (a) but as received from a position 1 metre behind (a) and 1 metre to the left of the common axis (centre line). There are *two* clear impulses, with the one arriving from the right-hand loudspeaker, later in time, and lower in level. The response from a central, mono loudspeaker, measured at position (b), would still be as shown in plot (a). It can be seen that off-axis, the propagation from a centrally panned stereo source and a central mono loudspeaker are very different.

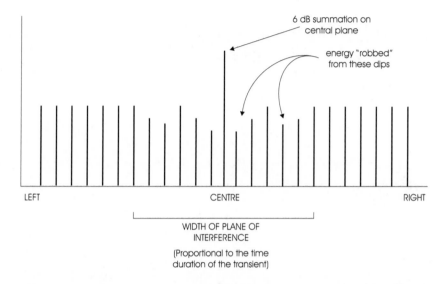

Figure 12.5 Transient superimposition. The 6 dB (double voltage) summation, on the central plane, of the overall power superimposition (3 dB) is gained at the expense of adjacent, destructive interference, elsewhere in the room

tence as separate bursts of energy, the performance of anechoic and reflective conditions differs only in that the reflective rooms will add an increasing number of reflected energy bursts to the environment, though of ever-decreasing energy, until they fade to oblivion in their multitudes. Perceptually, however, the reflected transients have the potential to mask the detail in any subsequent transients arriving before the reflexions from the previous ones have decayed to inaudibility. Anechoic spaces or acoustically dead monitoring environments do not suffer this limitation.

On 'steady state' signals, though, the situation can be very different between anechoic and reverberant rooms. As we have just discussed, in *anechoic conditions*, the interference pattern from the left and right loudspeakers would sum by 6 dB on the central plane, and for a distance either side of it which would be dependent upon wavelength. Away from the central plane the interference patterns would produce comb filtering, as shown in Figure 12.6, the nature of which would be position dependent. Off-axis, also, there would be additional low frequency power which was the result of the additional radiation due to the wholly constructive, less than half-wavelength mutual coupling. That this is not perceived on axis as an LF boost, but as an overall boost, can be considered to be a result of an overall directivity change when the two spaced drivers are operating together. Remember, we are still considering anechoic conditions here.

The width of the region of this uniform 6 dB, central plane pressure sum is very frequency dependent. At 20 kHz, the region of perfect summation would only be around 1 cm (half a wavelength), but at low frequencies would be many metres. At around 2 kHz, for a two-eared human being sat on the centre line, there would be an effective cancellation due to the spacing of the ears. Figure 12.7 shows how this is caused, as the path length differences are not

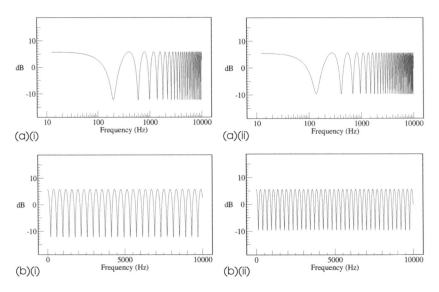

Figure 12.6 (a) Typical frequency response plots of a pair of loudspeakers producing a centrally panned image. Measurements taken at two different positions, off-axis, in an anechoic chamber. (b) As the plots in (a) but plotted on a linear frequency state. When viewed like this, on a linear frequency scale, the logic of the name comb-filtering becomes apparent.

the same from each loudspeaker to each ear. This is yet another mechanism by which a centrally panned image, from a pair of loudspeakers, differs in the way that it arrives at the ears compared to the arrival of the sound from a discrete, central loudspeaker, or an acoustic instrument. Conventional 'pan-potted' stereo is a psycho-acoustic trick, and is not a representation of any reality other than itself.

In the above paragraph, we have touched upon a situation where the *perception* of transients in *reflective* environments may be different to the perception of more steady state type signals as they are panned across the sound stage, even when listening directly on axis. The axial response of a stereo pair of perfect loudspeakers in a reverberant room was shown in Figure 12.1(b). Bearing in mind, that a reflective/reverberant room must be considered to be anechoic *until* the first reflexion arrives (by *definition*, it must be), then dependent on the room size and the length of the transient burst, the subjective perception of the transient could change over time from the anechoic, to the reverberant state, whilst continuous types of sounds (e.g. bass guitar notes) would be perceived more consistently. This leads to the conclusion that transient *vis-à-vis* steady state balances must be considered in accordance with the type of room in which they will be reproduced.

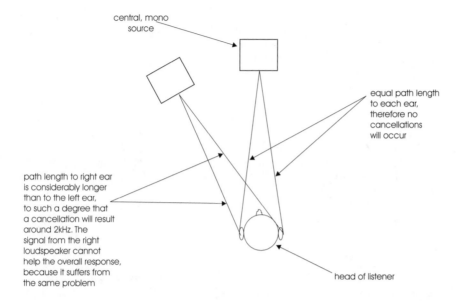

central, mono
source

equal path length
to each ear,
therefore no
cancellations
will occur

path length to right ear
is considerably longer
than to the left ear,
to such a degree that
a cancellation will result
around 2kHz. The
signal from the right
loudspeaker cannot
help the overall response,
because it suffers from
the same problem

head of listener

Figure 12.7 Path length anomalies for phantom central image

12.6 Performance differences between anechoic and reverberant spaces

If we locate a pair of loudspeakers in an anechoic chamber, then position them similarly in a reverberation chamber, we can, by use of similar drive signals, examine the two extremes of listening conditions. From the comparison of the two sets of results, we can thereby deduce some of the response irregularities encountered in many conventional rooms. We can also examine how a centrally panned stereo signal behaves in comparison to the same signal when reproduced via one loudspeaker, centrally positioned, both for different listening positions, and for different types of rooms.

In an anechoic chamber, a single loudspeaker with a flat, axial pressure amplitude response (frequency response) will deliver its output waveform to any position in the room which is within its cone of uniform, axial directivity. Elsewhere, the received waveform will be altered by the characteristics of the directivity of the loudspeaker. (Less high frequencies off axis, for example.) In all cases, however, a single pulse, emitted by the loudspeaker, will be *received* as a single pulse, anywhere in the chamber. Let us now consider two similar loudspeakers, placed around 3 metres apart, and each radiating an identical pulse to that emitted by the single loudspeaker just mentioned. To a person listening (or measuring) on axis, and equidistant from the two loudspeakers, the resulting sound would be a single pulse, identical to that received on axis from the single loudspeaker, though raised by 6 dB. However, at all points other than those which are equidistant from the two loudspeakers, two separate pulses would be received, as shown in Figure 12.4(b).

The precise perception of this effect varies from one person to another, but as the pulse from the nearer loudspeaker will arrive both ahead in time and higher in level, it will tend to swamp the lower level pulse, arriving later from the more distant loudspeaker. The tendency is to hear only the first pulse, and the apparent image source gets pulled towards the nearer loudspeaker (though the total loudness of the pulse may be perceived as being louder, due to a tendency of the ear to integrate the power of sounds arriving within 20 or 30 ms of each other). This is the Haas Effect, or Precedence Effect. There are time limits, however, beyond which the 'pulling power' of the first arrival breaks down. In outdoor locations, or in very large rooms, once a listening position is reached whereby the distance to the farther loudspeaker exceeds the distance to the nearer one by about 15 metres, which equates to about a 50 ms time difference, the later arrival becomes distinctly apparent as a separate event. This is why an effective stereo sound stage becomes impossible to support in very large venues, even though the relative distances to the two loudspeakers, and hence their relative SPLs, may not be all that different. There is only a very short window of arrival times in which the left and right signals can arrive if the stereo illusion is to be maintained. For most of the audience at a large concert, *stereo will fail to function*. Only the people close to the centre line between the loudspeakers will be lucky enough to hear the intended results. This is why central clusters are so popular, and so successful, with musical groups who play in large venues. Stereo causes confusion.

For 'steady state' signals, as opposed to transients, the centre line, axial response of a stereo pair of loudspeakers, generating identical signals, will be measured (at least with a single microphone) as having a similar response to a single loudspeaker, centrally positioned, and emitting the same signal, but receiving four times the power. To the human aural system, however, the response from the stereo pair will be modified by the inter-aural spacing (the width of the listener's head), as shown in Figure 12.7. Away from the central plane, the response will be comb-filtered at all but the lowest frequencies (where entirely constructive mutual coupling occurs) due to the different path lengths to each loudspeaker. The off centre line response of a centrally panned, phantom image, *will* not, and absolutely *cannot*, mimic the response from a centrally positioned, mono source, and will not even measure the same with a microphone. This fact is important in surround systems.

12.6.1 Reverberant complexity

If we now switch all of these considerations into a 'reverberant' environment, we will witness a new set of effects. The single, centrally positioned loudspeaker will, just as in the anechoic room, deliver its axial response to any point, within its axial directivity cone, forward of its main radiating axis. However, this response will only hold true until the arrival of the first reflected energy, after which, the response will be modified by the reflected energy field. The reverberant field will be of uniform intensity throughout the room, but the direct sound will fall by 6 dB for each doubling of distance beyond the near-field of the loudspeaker. So, as one moves further away from the source, the reflected energy will begin to become an increasing part of the overall, perceived sound. Hence the 'critical distance' which exists in reflective and 'reverberant' rooms, beyond which the room response will begin to dominate in the overall response. The additional reflected/reverberant energy will cause the signal to sound louder than in anechoic conditions, at an equivalent distance, but two other effects will also become apparent.

The first fact that falls out of these considerations is that, in anything other than more or less anechoic monitoring conditions, a 'steady state' type of signal, which receives modal support from the room reflexions in terms of its loudness, will be perceived differently to an impulsive, transient signal, which receives no such support. The loudness ratio of the perceived signals will be room dependent, *and* position dependent. In anechoic conditions, the transient *and* steady-state signals will fall off at the same 6 dB per doubling of distance from the source, and will thus maintain their relative balance, independent of time or position. The above effect in reflective environments is so very room dependent that no easy compromise monitoring conditions can be found. It can be a particular problem in larger listening rooms, such as cinemas. Mixes done in large, partially reverberant rooms will therefore tend to have different transient/steady-state levels than mixes done in smaller rooms. In rooms with relatively anechoic monitoring conditions, no such room size dependent variability would exist; or, if it did exist due to incomplete LF absorption, it would do so to a much lesser degree than in more lively conditions.

The second effect of switching our considerations to reverberant spaces is that, as the reverberant field is driven by the total sound output of the loudspeaker, its frequency balance will be that of the total radiated power. Typically, this will show a rise in low frequencies, as all directional loudspeakers with a flat axial response will need to generate extra power at low frequencies once directivity control is lost, and where the radiated power must be spread over a wider angle (360 degree below 300 Hz, or thereabouts). Therefore, for anything other than an omni-directional loudspeaker of uniform pressure amplitude response (a perfect loudspeaker?) the direct and reverberant sounds will not be perceived with the same timbral balance, so coloration of the sound will take place. This is why so much effort has been made, by many loudspeaker manufacturers, to produce loudspeakers with smoothly controlled off-axis radiating properties. These tend to produce reflexions and reverberation having frequency balances which, even if not uniform, at least are varying smoothly, so that at or beyond the critical distance less unnatural coloration is perceived by the listener.

12.7 Behaviour differences between phantom and real sources

In the case of a mono sound source, the coloration effect imparted by a normal room is problem enough, but when we generate our central stereo image as a phantom source, derived from an equal signal in each of the left and right loudspeakers of a stereo pair, an *absolutely different* and more complicated set of conditions exists. Even if we had a pair of 'perfect' omni-directional loudspeakers, the double impulse effect would still exist in all positions except those directly on a plane equidistant from the two loudspeakers. Only at very low frequencies, where the wavelengths were long in relation to the distance between the loudspeakers, would the pair begin to act as a single source and mimic the single, central loudspeaker in terms of the pattern of radiation off-axis, though they would *not* mimic the single, central loudspeaker in terms of frequency response: they would create a bass rise due to the mutual coupling.

In such conditions as above, if we were to install recording equipment and personnel, we would introduce reflexions and absorption, and here, our 'predictable' response would begin to break down. In truly reverberant conditions, the total power output of the two sources would be maintained in the reverberant field, irrespective of the fact that the interference between the two sources would radiate different, comb filtered responses in many different directions. However, once we introduce absorption, the absorbent areas will rob the reverberant field of the energy, either direct or reflected, which travels in the direction of the absorbent surfaces. As this will be non-uniform in frequency content, due to the interference products from the two sources, the overall response will thus become non-uniform. In the case of discrete reflexions, they would return reflected energy to the listener, once again, with a frequency balance which would be dependent upon the interference patterns radiated in the direction of the off-axis reflectors from the two sources. Again, this would be non-uniform in frequency content, so 'coloured' reflexions would be heard.

If we were to generate our central sound, not from the phantom source of a centrally panned signal produced by a pair of loudspeakers, but from a single central loudspeaker, then the absorber, at least in the case of the signal striking it directly, would, if uniformly absorbent with respect to frequency, absorb equal amounts of energy from all frequencies. It would thus tend to disturb the sound-field quantitatively, more than qualitatively: i.e., in loudness, rather than frequency balance. A similar situation applies in the case of reflexions, where a uniformly reflective surface in an off-axis location would reflect back to the listener a reflexion, perfectly uniform in frequency content, from the single, central loudspeaker source, but, a non-uniform, comb-filtered reflexion from a phantom source of a centrally panned signal from a stereo pair of loudspeakers.

So, even in a world with perfect reflectors, perfect absorbers, perfect diffusers and perfect omni-directional loudspeakers with perfect frequency responses, we still could not produce 'accurate' listening conditions for the centrally panned images in the sound stage created by two-loudspeaker stereo. It can thus be seen that a central image from a stereo pair of loudspeakers behaves very differently from a sound created by a discrete, central loudspeaker. Even if the radiated field from a single, central loudspeaker does not mimic that of the instrument which it is reproducing, at least the sources of all sounds are relatively co-located. This produces a strong argument in favour of three or five frontal loudspeakers, not only in terms of image stability when moving off-axis, but also from the point of view that fewer phantom sources and more discrete sources means that reflexion, absorption and diffusion will all be more uniform in frequency content. True, in real situations, the reflexion, absorption and diffusion will not all be uniform in their frequency response, but nonetheless, the less confusing interaction from multiple sources for single image positions means, at the very least, a more predictable set of starting conditions. Unfortunately, though, this still presumes perfect omni-directional, point-source loudspeakers with uniform frequency responses, and such devices do not exist. The message should, by now, be abundantly clear – loudspeaker reproduction of music is a very imperfect technology.

12.8 Limitations, exceptions and multi-channel considerations

Anyhow, even the *theoretical* concept of the perfect omni-directional loudspeaker breaks down badly when the question arises of where to put them. Within any room other than one which is either perfectly anechoic or perfectly reverberant (and they would have to be large for smooth LF responses) boundary reflexions from the rooms in which they were situated would produce an irregular frequency response. Perfectly reverberant conditions would be useless for monitoring, as the perception of most of the detail would be swamped by the reverberation. Anechoic conditions, in which fine detail is most readily perceived, allow no reflected energy, so the only sound heard by a listener is that which is received directly from the loudspeaker. Under such circumstances, however, there is no perceivable difference between a perfectly omni-directional loudspeaker, and one which radiates a uniform frequency balance on an axis pointing directly towards the listener. After all, in anechoic conditions, you only hear the sounds coming directly towards you. A uniform response for plus or minus 20 or 30 degrees off-axis allows for some movement about the central listening position, and this is quite easy to achieve in practice. So, omni-directional loudspeakers would do nothing here except waste power by radiating sound in unnecessary directions. It is a popular belief that the 'ultimate' loudspeaker would have to be omni-directional, but the above points show otherwise.

12.9 Multi-channel repercussions

The coupling situations which we have been considering can pose some very great problems for people trying to design electronic systems to 'fold-down' multi-channel mixes into stereo or mono, while still attempting to maintain the original musical balance. The desired fold-down would depend on the type of loudspeakers on which the music was mixed, as well as those on which it would be reproduced. It would also be greatly affected by the distance between the loudspeakers, both on mix down, and on reproduction, because the distances between the loudspeakers (and room boundaries) set the upper limits to which the wholly constructive low frequency mutual coupling can take effect. To make matters worse, the compatibility of mixes is also affected by the fact that the optimum fold-down must take into account the frequency dependence of the directionality of human hearing. The subjectively desirable level of a sound in a rear loudspeaker may be considered to be excessively loud, or bright, when reproduced from a frontal direction after fold-down. (Having been mixed with the back of the engineer's head pointing at the source.) Great care is required when considering what such fold-down systems can offer, and when assessing their claims about their degrees of compatibility. Compatibility with what? The electrical fold-down equation may differ very greatly from the purely acoustic, or the perceived, psycho-acoustic fold-down requirements. This harks back to Section 12.3, The pan-pot dilema. The acoustic nature of the listening environment will also add its own variables – the room coupling. Once again, though, relatively anechoic monitoring conditions would seem to offer the fewest complications for the electrical fold-down requirements, and, as the spaciousness would be in the surround, there would to little need for a room to add any more acoustic spaciousness if good surround monitoring or reproduction was required.

12.10 Summing up

So, once again, the search for the theoretically perfect loudspeaker is indeed a worthy aim, but in what sort of a room would it *behave* perfectly in stereo pairs? Only on the central plane in anechoic conditions, it would appear! In all other cases, the room will impart its influence upon the perceived response, and, even in the anechoic chamber, there are aspects of the weaknesses inherent in stereophonic reproduction which cannot precisely reconstruct the sound from a single source by means of a phantom image generated by two sources. The situation is that we have imperfect loudspeakers and imperfect rooms, trying to reproduce an imperfect principle via imperfect media. This is why it is the job of studio designers to get the best overall compromise out of any given set of circumstances. It is no easy task. Only by a very careful balancing of all the parameters can optimum end results be realistically hoped for, but parameters can only be balanced if they are understood, and many of the points being made here are not widely appreciated. The 'pot-luck', mix and match approach of the majority of project studios has almost no hope of avoiding the pitfalls, so the idiosyncratic monitoring conditions in many small studios comes as no surprise. Designers each have their own hierarchies of priorities, and from them we have quite a range of design concepts to choose from. The reason why no one principle of listening room design is universally accepted is because of the truly vast number of variables involved, of which the problems that have been discussed here form only a very small part. On the other hand, these concepts do have a great bearing on the *balance* of the compromises, especially when conditions for quality control monitoring are being considered. Unfortunately though, no designer can achieve perfection, because two-loudspeaker stereo, itself, is nothing but a *very* unstable illusion.

Appendix

After the construction of a significant number of control rooms, all using similar monitor systems, and all built to a similar philosophy, it was noted that a grouping had occurred in the nature of the porting of the vented boxes. The monitor systems used in these rooms had twin ports, the resonance peaking at around 20 Hz with both ports open. The studio owners were given the option of closing one or both ports, initially because it was thought that different musical styles may be more optimally monitored by different configurations – classical music perhaps favouring the sealed boxes. It was an idea which seemed to offer some flexibility.

However, after a period of about five years had elapsed since the first of the rooms had been built, a survey was made, in order to find out which studios had opted for which tuning. The results were very surprising. There was no correlation in terms of type of music recorded, analogue or digital recording systems, or any other of the expected reasons. The studios, in four different countries and recording very different music, had made the decision about the port tuning purely according to room size. The rooms below 60 m^3 all had both ports sealed, the rooms between 60 m^3 and 100 m^3 had one of the two ports sealed, and the rooms of over 100 m^3 had both ports open. It would seem that the rooms are loading the loudspeakers, and this relates to Dr Paul Darlington's questions, at a UK Institute of Acoustics conference in 1996, about the susceptibility of ports to room loading effects, as they are far from constant velocity sources, and may react to loading and coupling very differently to the loudspeaker drive units. This now begs the question as to whether it is possible to optimally design ported loudspeakers without full knowledge of the rooms in which they will be used.

In reference,[3] attention was drawn to work by Professor Frank Fahy at Southampton University, UK, showing how the effect of the introduction of a tuned resonator into a room can be perilously difficult to predict. The resonator, once introduced, changes the nature of the room itself, so the starting conditions (the room parameters) are no longer the same. In his aforementioned presentation to the Institute of Acoustics,[4] Paul Darlington put forward ideas for the active tuning of loudspeakers. The above findings may well suggest that there is a very real need for this.

References

1 Newell, P.R., Holland, K.R. and Hidley, T., 'Control Room Reverberation is Unwanted Noise', *Proceedings of the Institute of Acoustics, Reproduced Sound 10*, Vol. 16, Part 4, pp. 365–73 (1994)

2 Newell, P.R., 'The Non-Environment Control Room', *Studio Sound*, Vol. 33, No. 11, pp. 22–9, (1991)

3 Newell, P.R., *Studio Monitoring Design*, Focal Press, Oxford (1995)

4 Darlington, P. and Kragh, J.G., 'The Active Bass Reflex Enclosure', *Proceedings of the Institute of Acoustics, Reproduced Sound 12,* Vol. 18, Part 8, pp. 55–66 (1996)

5 Holland, K.R. and Newell, P.R., 'Loudspeakers, Mutual Coupling and Phantom Images in Rooms,' Presented at the AES 103rd Convention, New York, Pre-print No. 4581 (1997)

6 Holland, K.R. and Newell, P.R., 'Mutual Coupling in Multi-Channel Loudspeaker Systems,' *Proceedings of the Institute of Acoustics*, Vol. 19, Part 6, pp. 155–62 (1997)

Chapter 13

Phase, time and equalisation

The previous chapters have dealt quite extensively with the subjects of acquiring the cleanest signals, and using monitor systems through which their cleanliness can be perceived. Until now, though, we have not dwelt too much on the subject of precisely *why* we need to take such steps to 'simply' make a recording. What we now need to look at are the aspects of sounds which make such great demands on a recording process that we need to go to the lengths described to adequately monitor them.

13.1 Analogue, digital and sampling rates

Effectively, until the mid 1970s, music recording was an analogue process. To me, it still remains a remarkable fact that such great fidelity can be achieved from such 'crude' electro-magnetic and electro-mechanical recording devices. These performances were only achievable, though, after years of development, and the responses at the frequency limits were only achievable through the use of careful design compromises. In general, digital out-per-forms analogue at very low frequencies, but analogue can out-perform digital (certainly at 44.1 or 48 K sampling rates) at high frequencies.

Of course, if a characteristic analogue low frequency sound is wanted, as a part *of* a sound, then analogue could be considered to be better than digital for those purposes, but in my generalisations I am referring to the fidelity of recordings.

13.1.1 Phase and time responses

All analogue recording systems show phase disturbances in the lower octaves. These are created by the low frequency roll-offs that are encountered, usually between 15 and 30 Hz, in electro-mechanical and electro-magnetic transducers, such as tape heads and recording/ reproducing styli. The phase response disturbances, however, begin much further up the frequency band, perhaps to 80 Hz or so. Such effects are demanded by the physical laws which govern the processes, and usually cannot be compensated for in analogue systems. Nevertheless, they have become a part of the analogue sound, around which much recording has developed, and skilful exponents of the recording art have learned how to use them to advantage. In this way, analogue recording devices have become a part of many sound recording processes.

It is perhaps little wonder, therefore, that the simple substitution of a digital recorder in an analogue recording system has sometimes led to complaints about a 'digital sound' because a fundamental part of a balanced system has been replaced by something with which it has not evolved. Personally, I have no analogue/digital absolute preferences. I will use whatever devices are going to give me the sounds that I am looking for in the final mix. The phase accuracy and lack of any wow and flutter with digital recorders would make them my first choice for classical piano recordings, but I love the sound of analogue recordings for electric guitars.

In fact, in the case of piano recordings, there are two aspects of time accuracy which are important. Firstly, there is the lack of any noticeable variation in tape speed, the cyclic changes of which cause 'wow' when the rate of change is low, say 1 or 2 Hz, and 'flutter' when higher, say 15 Hz or so. Such disturbances can be very unpleasant on piano recordings, but can be benign, or sometimes even beneficial to guitars, whose players often introduce similar effects manually (whammy bars and string bendings). Secondly, though, there is the time response accuracy of the transient, or the 'leading edge' of a waveform. Distorted transients can severely affect the naturalness of the sound of an instrument, and this is where phase responses play a part.

13.1.2 Waveform

In the eighteenth century, Fourier, the French mathematician, showed that any sound could be broken down into, or constructed from, individual sine waves whose amplitude, frequency and phase were in a precise relationship. Any sound needs a time in which to exist, so it follows that its time response (i.e. the shape of its waveform) can be entirely constructed from its frequency and phase responses. Originally these things had to be calculated frequency by frequency, and were so unwieldy as to be impractical, so the concept remained largely theoretical until the advent of digital computers. The old concept was known as the Fourier Transform, but a new development of it, the *Fast Fourier Transform* (FFT), enabled computers to become very powerful devices for audio analysis. (Incidentally, an FFT is *not* simply a Fourier Transform done quickly, but is a development of it, designed for rapid calculation.) For many years, partly due to fundamental research on phase perception having been carried out in conditions which bore little relationship to music reproduction, and partly because the measurement of such things had been so difficult, it was largely accepted that phase responses had relatively little bearing on high fidelity sound reproduction, but now we know differently. In fact, it is largely the superiority of the phase accuracy of digital recording at low frequencies which gives digital its superior punch. On the other hand it is the superiority of the *high* frequency phase accuracy of analogue recording which gives it much of its characteristic 'sweetness'. Let us look at the low frequency performance first.

13.2 Amplitude and phase responses

Twenty Hz to 20 kHz is frequently considered to be an adequate bandwidth to cover the audible frequency range, but if we wished to reproduce a square wave with any degree of accuracy, say a 1 kHz square wave, then we would actually need a flat response from dc to infinity, free of all distortions. In reality, our combination of tactile *and* aural senses range from around 100 kHz down to the so-called 'weather frequencies' of maybe four cycles per *hour*, or around 0.001 Hz. A test signal which contains all of these frequencies is a step function, which can be generated in practice by applying a dc signal for a suitably long period of time. Theoretically, it would have to remain on for an infinite length of time, but in practice this is not necessary for our purposes. A step function can be thought of as one half of a square wave with an 'on' time long enough for our measurements. It contains *all* frequencies in a fixed relationship of amplitude and phase. Connecting a battery to the terminals of a loudspeaker and leaving it there produces a useful step function, and is an effective way of producing a very revealing test of loudspeaker performance. To save the drain on the battery, and to drive the loudspeaker from a more realistic source, one would usually perform such a test by connecting a small dc signal to the input of a wide bandwidth dc amplifier, which would in turn be connected to the loudspeaker. From this single pulse, or preferably from the averaging of a series of them by Fast Fourier analysis, both the frequency response and phase response of the loudspeaker and amplifier can be measured to a very high degree of accuracy.

Figure 5.1 showed the step responses (the output waveforms) of several loudspeakers, all of which are used as studio monitor loudspeakers. The output waveforms are their attempts to follow the electrical input signal shown at the top of the figure. Although all of these loudspeakers have 'frequency responses' or more correctly 'pressure amplitude responses' which would generally be considered good, their phase responses, even on axis, vary to such a degree that the time response waveforms shown in Figure 5.1 are so very different. It is the convolution, the interaction, of amplitude *and* phase which produce the *time* response, or in more simple language, the waveforms. All of the loudspeakers shown in the above figure are very high quality units, but they all sound different. The difference in sound would not be expected from a visual assessment of the frequency responses alone, two of which are shown in Figure 13.1, but from examination of the time waveforms of Figure 5.1, I do not think that the differences would be quite so unexpected.

Incidentally, the reason why none of the responses stay at the constant level of the input signal is that whilst the loudspeaker cone may remain displaced by the dc input signal, the measurement of the output would only remain constant if the loudspeaker and the measuring microphone were in totally sealed boxes (a sealed loudspeaker cabinet and an air-tight listening room). The displacement pressure caused by the one direction of movement of the loudspeaker cone soon leaks out through the doors or ventilation system of the room. Moreover, if the loudspeaker cabinet is a bass reflex type, then the front to back pressure differential of the loudspeaker, caused by the displacement of the cone, soon equalises through the tuning port. However, this is an effect which occurs over a short time, and the ultra low frequency response is still desirable in order to prevent the distortion of the waveform on the attack of the transient. Remember, a low frequency roll-off in amplitude produces phase response errors, which then produce transient distortion.

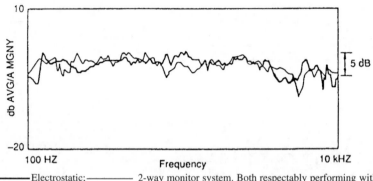

——————— Electrostatic; ——————— 2-way monitor system. Both respectably performing within ± 2½ dB over most of their range

Figure 13.1 In Figure 5.1 we saw the step function responses of four loudspeakers (on more or less similar time scales) all of high repute, compared to the electrical input signal. The slope of the response tails is due to their low frequency response limits. Sonically (a) and (b) in Figure 5.1 are reasonably similar, whilst (c) and (d) sound characteristically different. The step responses suggest that this should be the case. The plot here however shows the frequency responses (pressure amplitude response) of (b) and (d) from Figure 5.1. From these frequency responses alone, it is by no means obvious that the two should sound significantly different, nor which of the two should sound the most realistic

From the waveforms in Figure 5.1, we can largely see only low frequency information, as the high frequency information is mainly concentrated within the leading edges. These sorts of response differences often used to be masked by the low frequency response irregularities of analogue recording devices, but digital recordings have laid them bare. As all analogue cutter heads, pick-ups and tape heads had their characteristic time responses, which were not too dissimilar from the difference between the loudspeaker responses, it is not too surprising that it was only when digital recorders became widely available that any big work began on improving phase accuracy. If any reader has an oscilloscope handy, then they can have great fun (or lose their hair!) looking at the attempt of analogue recorders to reproduce accurate square waves. The waveform distortion can be startling.

Some sounds are quite robust in terms of their tolerance of such phase distortions without seemingly changing their characteristics, but others can be particularly sensitive to them. Unfortunately, only experience can give good advance knowledge of what to expect, but even then there can be big differences in the amount of sensitivity which different people exhibit to the differences, and even more in the amount of importance which they attach to them in terms of their enjoyment. This last comment leads us to a point whereby we may be able to create 28 bit recording systems, sampling at around 500 kHz, which completely out-perform analogue recorders in *all* areas of accuracy, yet which few people would buy. Some people may simply prefer the analogue sound, whilst others may perceive so little improvement over 16 bit linear/48 kHz sampling, that they simply would not be prepared to pay the extra price for the 'super system'. On the other hand, there could be other people who would 'kill' for such machines.

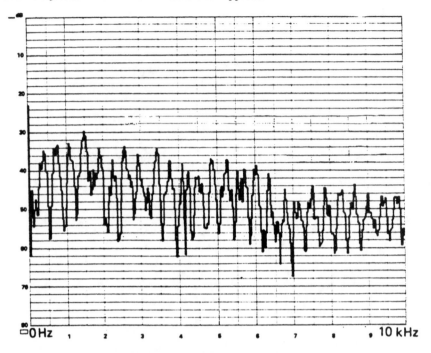

Figure 13.2 Comb filtering. The averaged power spectrum of a signal with one discrete reflex-ion. Comb filtering is revealed clearly in a linear (as opposed to a log) plot, where the regular nature of the reflexion-produced disturbances can clearly be seen. In the instance shown here, the additional path length of the reflected signal over the direct signal was just under one metre, producing comb filtering with dips at a constant frequency spacing of just under 400 Hz

13.3 Waveform perception

The presentations in Figure 5.1 were on a time scale of around 5 ms. There is an argument that the ear tends to integrate packages of energy about one hundred times per second (every 20 ms) so the differences shown in the figure should not be audible if they contain the same amount of energy at each fre-quency. On musical signals, however, there is much evidence to show that such waveform differences *are* audible. However, with waveform differences on a longer scale, there is little argument that the perceived differences can be gross.

Figure 13.2 shows the 'comb filtering' effect created by a hard, reflective surface, close by one side of a loudspeaker. The plot is on a linear frequency scale, unlike the usual logarithmic presentations, to emphasise the periodicity of the disturbance. Ears can be very sensitive to the recognition of patterns, so this type of coloration can be readily noticeable. Indeed, it is used as an effect when single fast repeats are used from electronic signal processors. When judging such sounds then, it is perhaps better for the loudspeaker room response to be free of them, hence the concepts of monitoring discussed in Chapter 11.

Figure 13.3 is largely self-explanatory – loudspeakers alone do not define

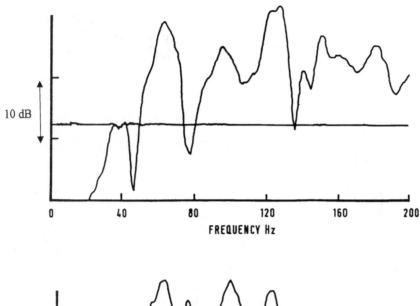

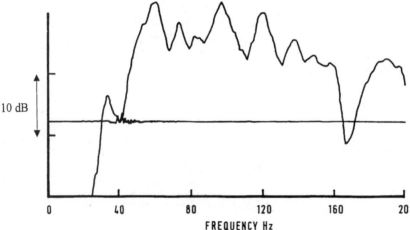

Figure 13.3 Response disturbances caused by different room positions. Low frequency response plots of one loudspeaker in one control room but in two different positions

a monitoring system response. Figure 13.4 requires a little further explanation, however. It shows how the path length, and hence the arrival delays, will bear a different relationship to each other when listening off axis. If the response at point A is designed to produce a synchronised arrival of sound from each of the three loudspeakers, then clearly, at point B, the arrival times will be spread out. The figure represents a two-way loudspeaker with a tweeter sandwiched between two woofers, but it could equally represent a simple two-driver, two-way system by ignoring either of the woofers. Whilst sat in a chair behind a mixing console and rolling from side to side, the

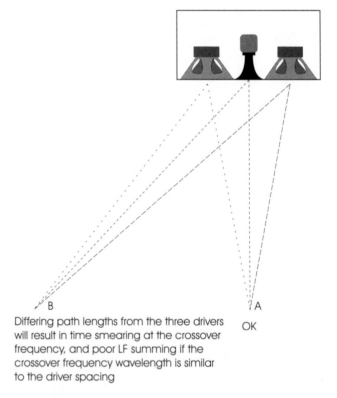

Differing path lengths from the three drivers will result in time smearing at the crossover frequency, and poor LF summing if the crossover frequency wavelength is similar to the driver spacing

Figure 13.4 Driver position considerations. Horizontal mounting of the monitor system. At position A all is well with symmetrical positioning allowing equal arrival times from the three drivers to the ear. For position B, however, the path length, and hence the arrival time of the signal from the left-hand bass driver, is shorter than that from the horn, whereas that from the right-hand driver is longer. A flat frequency response cannot exist at both positions

response of one loudspeaker system, horizontally positioned as in the figure, will give a continually varying response, even with a mono signal. Heaven *only* knows what will happen to a phantom central image from a stereo pair.

If the loudspeaker shown in Figure 13.4 was mounted vertically, with all three drivers in line, then horizontal movements behind the mixing console would *not* produce any differences in the path lengths to the loudspeakers. Yes, I *know* that Yamaha print the labels on NS10Ms in such a way that leads users to lie them on their longer sides, but this is an extreme nonsense. Serious monitor manufacturers such as ATC, Quested, Tannoy, Genelec, KRK and the like would never dream of doing anything similar. In fact, Figure 13.5 was taken from a Quested catalogue, and makes absolute sense.

The above fact causes much time-wasting in many studios when people are listening off axis; even only slightly. In such cases, switching between the large and small monitor systems may produce an acceptable correspondence in sounds when listening at their point of convergence, but from even only 30° off the axis of the smaller pair, switching between the two systems will produce an appalling mismatch in sound character. This can lead to much

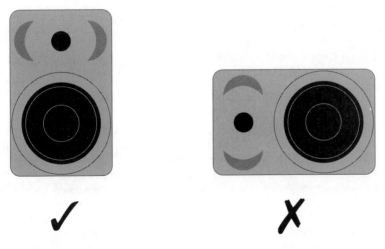

Figure 13.5 Correct use of small loudspeakers. More and more loudspeaker manufacturers, such as Tannoy, Quested and Genelec, are now publishing drawings such as these in their manuals for small loudspeakers, instructing their users to keep the drivers on a common vertical plane. It matters naught that there are thousands of studios using such loudspeakers on their sides – the practice is WRONG

misinterpretation of equalisation by producers or band members listening from different parts of a control room. The tendency here is that the better the control room, the greater will be the perceived differences.

Hopefully, it can be seen from the previous paragraphs that electro-acoustic, electro-magnetic and electro-mechanical systems all have conversion difficulties which can affect the perceived tonality of a sound, and that is why so much stress has been made in this book about the need for monitoring accuracy, which is the *only* means by which many problems can be assessed.

13.4 Equalisers

It is not only the transduction systems which can introduce phase errors, though. Equalisers are notorious for the introduction of phase anomalies. Indeed, much of the difference in sound between equalisers, which *appear* to be set to the same curve, can often be tracked down to phase differences between the different implementations of filter design. In many cases, when used for sound effects, the ringing in the filters and the characteristic sounds which they produce can be considered to be desirable assets. The real absurdity comes in, however, when we try to use equalisers on monitor systems.

13.4.1 Monitor equalisation

It was undoubtedly an aberration of the 1970s which led to the absurdity of monitor equalisation still being considered by many inexperienced people to be a 'must' in professional monitor systems. In fact, nothing could be further from the truth. Any studio which has 'corrective' equalisation on its monitors is demonstrating that:

1 It has an inappropriately designed control room.
2 It has inadequately designed monitor loudspeaker systems.
3 They do not know what they are doing.
4 Any combination of the above.

The only, somewhat rare, exceptions to these comments are where the monitor equalisers are in use to gently modify a minimum-phase problem.

('Minimum-phase' problems, remember, are ones in which the amplitude corrections will automatically tend to improve the phase response, and hence the time (waveform) response.) Such minimum-phase situations would be to correct for a gently falling or rising loudspeaker response, or to achieve a 'house curve', when it was considered that a monitor system was over-bright, or bass-light, for example, or where a room loading effect, as discussed in Chapter 5, created, for example, a boost due to flush-mounting a loudspeaker which had been designed for stand-mounting. Many of the Genelec and Quested ranges of loudspeakers have equalisation switches, *and* correct instructions, for use in just such circumstances. It is a growing trend.

However, these are not the uses for which monitor equalisation is usually, and quite inappropriately applied. The usual reason for the use of monitor equalisation is to fix 'room problems', 'standing wave problems' or to 'voice the room' in the old jargon. Conventional monitor equalisation *absolutely* cannot be used for these purposes. Room reflexion and resonance problems are created when energy returns to the listening position after having travelled a longer path than the direct sound. There is therefore a delay between the direct signal arriving at the ears, and the 'problem' reflected energy. There will therefore also be a phase difference between the direct and reflected energy which is entirely dependent upon frequency. If every frequency in a reflexion has a different phase relationship to the direct signal, and what is more if the phase differences are different for every reflexion and every room position, then no conventional equaliser can correct the phase as it corrects the *average* amplitude. Clearly it would have to be an average amplitude correction, because no conventional equaliser could hope to follow the actual inverse of the response shown in Figure 13.2. Only filters such as those described in Section 13.5 can do this, but even they have their limitations.

If a room has a resonance at 80 Hz, and an equalisation of −5 dB brings the pink noise response to a flat reading on a spectrum analyser, then the equaliser may have corrected the averaged steady state amplitude response. However, as described in Chapter 12, all rooms, by definition, are anechoic until the first reflexion arrives. As a monitor equaliser acts on the drive signal to the loudspeakers, then if we make a 5 dB cut in the monitor response at 80 Hz to 'correct' the room resonance, then we will also cut 5 dB at 80 Hz from the transient signals, which arrive at the ear *before* any resonance is heard. The equaliser will therefore distort the amplitude, the phase, *and* the time response (waveform) of the transient. It will also, therefore, distort similarly all the reflexions of that transient which do not arrive via the resonant pathway that created the 80 Hz hump.

Steady state sounds, such as a bowed cello, may well benefit from the adjustment, but transient sounds, such as drum beats, or the attack of an acoustic guitar, may be ruined. Problems caused by the arrival of reflected energy, and which thus have a delay relative to the direct sound, are 'non minimum-phase' problems, and they *most definitely cannot* be corrected by conventional monitor equalisation.

Remember; because different path lengths, and hence delays, are involved, the effects of a non minimum-phase problem will be position dependent, so they cannot be corrected by a simple change in the drive signal to the loudspeakers. Low frequency loading boosts or cuts, such as those encountered by the flush mounting of free-field monitors, *are* minimum-phase problems, as no signal delays are involved, so they are *not* position dependent, and they *can* be equalised. I cannot over-emphasise the importance of understanding this difference, because the lack of understanding has been the ruin of countless monitor systems.

The same comments apply to the use of equalisers in clubs, discotheques and live concert venues. Conventional equalisers cannot correct 'room problems' without the distortion of the transient waveforms and the creation of position dependent anomalies. What is more, if, under any circumstances, you see an attempt at a 'correction' which involves the ups and downs of adjacent filter bands, they you *know*, for sure, that you are witnessing a travesty, no matter what the reading on any real-time analyser.

13.5 Adaptive digital signal processing

Despite all of the problems in the electro-acoustic systems, technology is advancing, and as it moves on, some of these very real problems will be overcome. Adaptive Digital Signal Processing is beginning to feature in many people's thinking. Localised control of the sound field for a small area within a room can be achieved by using digitally adaptive filters and modelling delays. By such means, control signals can be added to the audio signal, to provide the inverse of both the loudspeaker deficiencies *and* the reflexion/time response problems of the rooms. The delays are necessary in the signal path in order that acausal correction signals (where the correction signal effectively needs to be applied marginally before the actual event takes place) can be made to anticipate and nullify the unwanted effects. Before seriously widespread use can be expected, however, more work needs to be done on the subjective audibility of some of the correction processes, such as the pre-echoes which are produced as an inevitable result of the processing. Costs will reduce as the more powerful micro-processors are developed in greater quantity.

By means of such systems, in a designated listening area, remarkably accurate response measurements can be made, though at almost all places outside of the designated area the response will actually be worse than before processing. Binaural separation can already be achieved to such a degree that the right ear can be totally unaware of the signal from the left loudspeaker, and the left ear can be deaf to the right loudspeaker, as in headphone listening. The power of control is great, but as all of the control signals need to be superimposed on the audio signal, the entire amplifier and loudspeaker system must have sufficient headroom to take the much greater signal peaks. The more flat the overall system response is to begin with, the lower will be the relative level of the necessary control signals. In a well-controlled room with a highly uniform monitor system, excellent results can now be achieved; though somewhat ironically here, the better the electro-acoustic system is to begin with, the less it will probably *need* the active processing. There are, however, some loudspeaker system response problems, such as some of the amplitude/phase characteristics at crossover points, which can be corrected by no other known means than by such adaptive filtering. Digital control may therefore eventually become a final embellishment on already excellent systems, rather than a cure-all for inferior systems. Clearly, though, the frequency range of any digital system which is used for monitor correction must be wider than that of any item in the recording chain. *There would seem to be little point in trying to monitor a signal from a recorder with a 96 kHz sampling rate, through a monitor system using signal processing with only a 48 kHz sampling rate.*

13.6 Why bother?

What one tends to notice between top end and 'promestic' equalisers and effects is that the top end equipment tends to be more subtle in what it does. Clearly, when recording first rate musicians with first rate instruments, in good rooms, one would rarely want to adjust their sounds with 'sledge-hammers'. Conversely, when searching for new ideas when recording less elevated performances for demonstration purposes, a little extra 'help' may be required. What is more, though, subtle sounds require subtle monitoring; and that is something which can be pitifully lacking in may project studios.

I have known a number of effects units which were designed for the semi-professional market as cheaper alternatives to 'classic' professional units. Some of them failed to sell to their intended markets in large quantities because their effects were too subtle to be noticed in poor surroundings or through less transparent signal paths. They were subsequently snapped up, however, by the top end studios as cheaper alternatives to the classic units, some of which were no longer in production. This is one example of the tendency for the studios with less experienced staff to buy 'names' rather than performance, whereas the top studios tend to buy performance, irrespective of the name. Once again, this highlights a weakness which marketing departments exploit ruthlessly. They can even encourage insecurity in order to stimulate the perceived need for items which are thought to provide security. (An old trick of the 'protection' racketeers.)

On the subject of names, I recall doing one recording for a company who owned a small recording studio. It was to be quite an 'important' jazz recording, which they had great hopes for. The studio's own microphone selection was hardly adequate for what we wanted to achieve, but the studio owner told me that he could borrow some microphones called 'Scopes', or something similar. When they arrived, it turned out that he had a friend in television who had lent him a full set of top of the range Schoeps microphones. I set up the recording, made the sound checks, and we were all ready to begin when the studio owner arrived, with his face beaming, saying 'Let's go for an early lunch, because I've arranged to borrow some Neumann U87s at 2 o'clock'. He was really feeling proud of himself, and could not understand my startled expression and wrinkled nose. Now, I would not have turned my nose up if the studio had offered me U87s at the outset, but I was not about to swap Rolls-Royces for Jaguars.

Two days later, and after recording some wonderful sounds, the studio owner was still asking me 'Are you sure that it would not have been wiser not to use the Neumans? Would we not be getting better sounds?' He simply refused to believe his ears (the sound that we had was excellent) because he was brainwashed by the (fully deserved) reputation of Neumann, but had never heard of Schoeps. Now this raises another interesting point. If the Neumann aus Schoeps microphones are both so commendably flat in their responses, then how come they sound so different? There is clearly more to it than the frequency response. As you cannot equalise one microphone to sound like another, you cannot do so with loudspeakers, either.

What has been written about in this chapter is important in terms of sound quality, at least if the word 'quality' is used synonymously with fidelity. These problems will not go away if you ignore them; they will return to haunt you. I know the problems and respect their influences over what I am doing. It is my ears, and *only* my ears, that tell me whether or not any equipment or conditions are suitable for high quality recording, or not, but only experience has given me the experience to use them. Of course, I did have some good teachers, as well. It is interesting to note, though, that there often occur occasions when I am encountered doing live recordings. On some of these occasions, when well-known, successful recording engineers find me whilst working, the conversations tend towards 'Hi Philip, how are things going? They're playing well. The pianist is playing a little mechanically, though.' The big contrast is when inexperienced engineers approach me. 'Hi Philip, Oh, I see you're only using a Mackie, I expected that you would be using something a little better.' 'Why, what is wrong with the sound?' I reply. 'Er, ... well, ... no, ... the sounds are great,' they say. 'Then what is wrong with the Mackie?' I ask. 'Er, ... what microphone is on the bass?' 'Why,' I ask. The implication, usually, is that if I get a great bass sound with this microphone, then they too will get a great bass sound when using a similar microphone, even on a different instrument, in a different room, and with a different bassist. In fact, whilst on this subject, I remember seeing an 'engineer' come into a studio to begin a session, and pre-set the console equalisation according to a list that he had made up of equalisation settings used by famous engineers on famous recordings. I was speechless. He was not listening; only looking at a list which related to an entirely different set of circumstances.

Decisions of a truly professional quality can only be made by a combination of careful listening and an understanding of the processes of sound recording and reproduction. It is hoped that much of what has been referred to in this chapter will help to develop that understanding.

Chapter 14

Computers in control rooms

There are relatively few control rooms, these days, in which the use of computers has not made an impact. This is true in the cases of professional, semi-professional and domestic use, though the reasons for their widespread use is not necessarily the same in all cases. Quite simply, computers and digital processing make some things possible which could not be achieved in the analogue domain, such as real-time de-noising. They also make some things affordable which would not be economical to produce in an analogue form, such as the Yamaha 01V mixing console.

In this age of so much 'virtual' everything, it is little wonder that the software packages which turn almost any home computer into a virtual recording studio are so popular. These systems have made possible the single room, single operator studios which seemingly have all the facilities of a full professional facility, although their practicality of operation can be very limited, dependent upon the circumstances. The current tendencies in the film-dubbing world, for example, seem to be to leave every option open until the last minute. Everything tends to be stored and programmed pending final decisions, when, one by one, things can be called up, and used as building blocks to create a final reality from the director's dreams. Computers are in widespread use for this type of work, and are immensely useful. This is a far cry, however, from the needs of recording an orchestra of very expensive musicians, where there is a requirement for virtually instantaneous control of nearly every function. In situations such as this, one control, one function and rapid manual access is super-important. Dialling up menus and messing with the mouse is out of the question.

14.1 Domestic studio advances

By its very nature, home recording tends to be a rather leisurely process. In many cases, much of the enjoyment is gained from being able to explore so many possibilities, and to do for affordable prices what only the large professional set-ups could do before. Software packages can often turn a PC or a Power Mac into a whole virtual studio for the price of a single, high quality, professional effects unit. The drawback is often that it may take 20 minutes of 'fiddling' to achieve what the professionals could do in 10 seconds. Time pressures and costs are often the arbiters in such circumstances.

Small recording consoles such as the aforementioned Yamaha 01V are astounding value for money in terms of what possibilities they offer for the price. Despite the fact that more or less everything must be dialled up, and very few things can be done 'at the touch of a button', these consoles may well have no analogue counterpart. Let us imagine a musician, who wants to record at home, but has a limited budget and limited space. To imitate the 01V by analogue components would likely fill the room, and may actually cost as much as the house. Then, of course, there would be the cost of the air-conditioning to get rid of all the heat.

If the studio was only going to be of use if it had a sufficiently wide range of capabilities, and rapid real-time control was not needed, then the very question of the studio's existence, or not, may only be answerable in the affirmative by the existence of such consoles as the 01V. There is absolutely no doubt that digital processing and computer control have opened up a huge market. They have enabled an enormous number of domestic studios to come into existence, many of which would never have been sufficiently inspiring for their owners to build them if they could only offer the facilities of whatever could be bought in the analogue domain for the same price.

14.2 Professional expansion of capabilities

At the top end of the recording industry, digital equipment is usually expensive. Work has to be able to be done quickly and with the minimum of breakdowns. Computer systems can control instruments, machinery, mixing console automation, and many other things, but unlike at the domestic end of things, separate computers are usually used for each purpose, with the operating systems being asked to do perhaps only one job. The tendency is therefore towards the use of much more hardware. Crashes are much less likely when the systems are not also used to run the family accounts and the children's games. What is more, should a crash occur, then it is likely only to take out some single function of the studio, and not the entire operation.

In many cases, top line studios use computers to offer facilities which analogue systems cannot provide. Computer-based editing can achieve results which were only ever dreamed of with analogue systems. Twenty years ago, we edited the analogue 2 inch and ¼ inch tapes with razor blades, and could sometimes make a hundred, or more, edits for a classical recording or a 'concept' album. With the required skill, these could be done equally as well as any digital device can achieve, but they could also be fraught with difficulties or untraceable noises.

Magnetised razor blades were often blamed, but in many cases no amount of de-magnetising would produce silent edits. Tape manufacturers sometimes admitted to poor bulk erasing during manufacturing, which could produce low frequency variations that the split erase heads on many '2-track' machines could not erase from the guard band, between the tacks. When cutting on a peak of such a signal, the 'thump' could leak across into the adjacent head gaps. Nevertheless, even on 'stereo' machines (as opposed to 2-track) which had mono, full track erase heads (which precluded the recording of each track at different times) we could still get 'clicks' at the edit points. This was often due to the relative phases of the instrumental signals being such that no common zero-crossing point could be found on the left and right channels simultaneously. The edit would cause some jagged waveform to be produced when the two sections were spliced together, and this would produce a 'click' or a 'thump'. Computer based editing systems can now provide mouse-controlled virtual pencils and rubbers, to erase the offending click signals and re-draw a smooth waveform. When I first saw these, I was astounded. This was science-fiction come true. The answer to a dream. For the uninitiated, Figure 14.1 shows the principle.

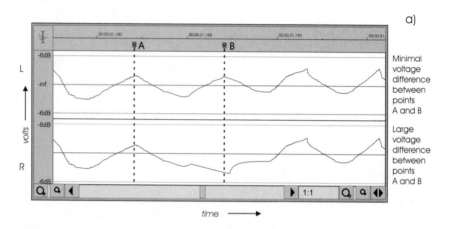

Figure 14.1 (a–c) Edit-point waveform correction. (a) Removing the section **A–B** would produce a smooth continuation on the left channel, but an abrupt jump in level on the right channel. Points **A** and **B** both represent small positive voltages on the left channel, but **A** represents a small positive voltage on the right channel, with **B** representing a large negative voltage. When the two points, **A** and **B**, are joined after removing the section between them, the sudden switch in voltage will produce an audible click on the right-hand channel.

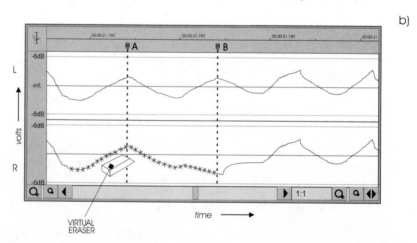

Figure 14.1 *cont.* (b) By means of using a software controlled 'virtual eraser', and moving it around by mouse, the crossed out section of the waveform can be removed, along with a short lead-up section

c)

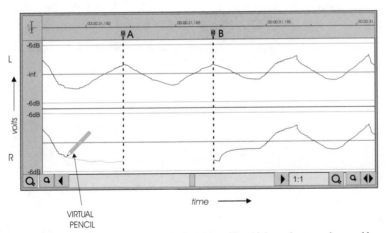

VIRTUAL
PENCIL

Figure 14.1 *cont.* (c) Finally, by means of 'virtual pencil', which can be moved around by mouse, a new section of waveform can be drawn in (shown dotted). When **A** and **B** are then joined together, no voltage change will occur on either channel at the edit point, and no click will result

14.2.1 A rare edit

Track synchronisation, correction of pitching errors, correction of timing errors and 'test the overdub without losing the original' are all products of computer-based technology. These are more things which were not available without computer technology. I remember one instance when Tom Newman was engineering a session for Dick Heckstall-Smith with Jon Heismann. The 16-track 2 inch tape needed editing, but there was one track which needed to decay naturally, across the edit, for about three-quarters of a second. Tom measured the tape with a ruler, decided where the appropriate track was, and cut the tape on an editing block across the other fifteen tracks. This left the tape joined by about one-eighth of an inch (3 mm) at the required track. He then cut two parallel lines in the tape, either side of the wanted track, for about 10 or 11 inches (we were running at 15 inches per second) then completed the separation of the tape. He then had to cut a corresponding slot in the section to which it was to be joined, pieced the two wanted ends together, male and female, covered the join with ¼ inch splicing tape, and used ½ inch splicing tape across the tape in the region of the main splice.

I witnessed this, at The Manor Studio, with my own eyes. I witnessed the prior discussions, also. 'Tom, you can't do this,' I said. I then went into the house, as I could not bear to watch. 'He is nuts,' I commented to anybody prepared to listen to me. About half an hour later, the faces of the musicians coming out of the studio could only mean one thing. The edit had worked. On Dick Heckstall-Smith's album 'A Story Ended' there are credits to Tom 'Risky-Edits' Newman – Courage beyond the call of duty, there'. I had to take my hat off to him. If you can bear the strain, see Figure 14.2. Although clearly not impossible, the above system of editing is not recommended. Nowadays, such daunting problems are easily solved by computerised editing, without any thought of the prior difficulties in achieving such tasks in the analogue domain. De-clicking, de-scratching, de-popping and de-noising programs

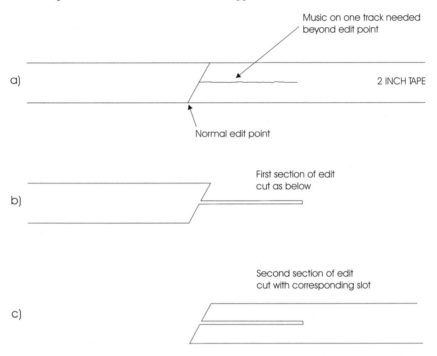

Figure 14.2 A risky edit. Sections (b) and (c) were overlayed and spliced together with self-adhesive ¼ inch splicing tape. Half-inch splicing tape was used along the diagonal cut

also belong to the world of computers – they have no analogue counterparts. However, a good set of these boxes will cost much more than most entire domestic studio set-ups.

14.2.2 Digital control of analogue

Microprocessor control of analogue circuitry is another role for computers in high quality music studios. Consoles such as those manufactured by Euphonix are good examples. In these designs, all the analogue circuitry is housed in 'audio towers', which are free-standing racks, normally kept out of the control rooms. They are noisy, and produce a lot of heat, neither of which is desirable in *any* control room, and they are out of the question in professional installations. The console 'work surface' is a digital control panel, with motorised faders. It is cool and slim – ideal properties for control room use – but they *are* quite large. This is necessary because, for professional use, speed of operation is important. Sufficient controls need to be accessible at any one time, even if they are multi-function, for the console to be configured such that all of the operations that need to be done 'instantly' can be done with a minimum of delay. If the lead vocal is getting very close to overload on a 'one and only' performance there is no time to dial up a menu, run a mouse around, press the appropriate keys, and … too late; the track over-loaded. These things cannot be achieved with small control surfaces and only a few knobs or buttons.

14.3 Virtual mixers and hard disc recorders

There has been rapid growth in the number of studios using computer-based mixing systems and hard disc recording. In the world of project studios, they have become both an asset and a liability. The professionals use such equipment where it is both effective *and* practical, and use other systems whenever necessary. The domestic users are usually very pleased to have the capacity for working to such high standards at an affordable price, and happily accept the restrictions which the systems may impose. The problems exist in the middle order, where the attractive prices and seemingly boundless capabilities can lead to some quite complicated systems which seem to rival the facilities of many major studios. The question is, however, can they offer it all at the same time? The hardware oriented professional studios can, but the middle order usually cannot, and this can lead to frustration and bad working practices.

I recall finishing the acoustic construction of a studio in 1996, into which had been installed an excellent PC based multi-track recording system, with hard drives. A band had been arranged to do some trial recordings, and when I visited during the first week of use, the band were intrigued to know how a 'known' engineer would do things, and they absolutely insisted that I recorded a track for them. The situation was impossible to get out of, so I sat behind the mixing console and we started running through the number. We ran into problems within the first 15 minutes, and the situation was one which I would not wish to repeat.

I was trying to record a rock band, with a five-piece rhythm section. Including guide vocals, I needed about 20 channels and about 16 tracks. We had sufficient virtual tracks, and sufficient channels on the mixing desk. We also had sufficient sub-groups, but we were short of analogue to digital converters (A to Ds) to put everything down together. As a sometime producer, that was the way that I wanted to record the band. The band responded well to my wishes, but we could not do what I wanted to do. The studio manager and engineer suggested that I should record the instruments either separately, or in pairs, to which I replied that I was trying to record a rock band, and that rock bands usually need to play together. The band's eyes lit up. Somebody actually wanted to record them the way that they wanted to play.

Their enthusiasm was uncontrollable, and the manager and engineer tried to explain how they *could* be recorded, but the genie was already out of the bottle. The whole thing turned into a heated discussion with the band insisting to the studio people that what I was describing was how they always *thought* that they should be recorded. Previously, however, studio staff had always managed to convince them that that was not how recordings were made, at least not these days. I *had* to side with the band, but I was desperately trying not to hurt the feelings of the studio people, or to embarrass them. How could I put it diplomatically to them that the recording system which they had chosen, and which had been so well reviewed in the magazines, was actually not suitable for all the types of recording which they had envisaged. They were insisting, to the band, that they *could* be recorded, but they would have to record in a way which the system would handle. The band were on the point of accepting this, and in fact they would have done, if I had not opened my mouth, again, to say that the job of a studio was to record musicians, and that it was not the job of musicians to play to the operating systems of a studio. I really did not want to deflate the pride of the new studio owners, but neither could I see a band used as computer fodder. It was not one of my happiest days, but the studio owners took it philosophically, although only perhaps by deference to my experience. Had I not been there, the argument surely would have swung the other way. Ultimately, although I recorded nothing that day, the band were very encouraged to discover that, at the higher levels, at least, the studio was the slave to the musicians, and not vice versa, as frequently seems to be the case in the middle order.

All too often, in mid-priced studios, the technology rules. In fact, the power and influence of the technology has become so intimidating that it can take quite some courage, or experience, to challenge it. Experience, unfortunately, is often all too lacking in these studios, but there again, big studios with experienced staff cost a lot of money, and one tends to get what one pays for. It is strange, though, how the technology, and the marketing of it, leads the unsuspecting into making decisions which are wholly inappropriate to their needs. It is not the job of the virtual studio manufacturers to tell studios how they should record musicians.

14.4 The digital lure

I recently turned down a job (or rather, lost it by default) to design a complex of rooms in a college which was intending to provide training for recording engineers. They were also intending to issue certificates of qualification to the successful students. My reluctance to participate was spawned by the attitude of the directors, for whom the marketing of the course was more important than the standards of qualifications.

'The Future is Digital' was to be their catchphrase, but they were a little surprised at my aggressive demolition of their philosophies, such as the use of ridiculously small control rooms, and the cursory provision of an analogue room which was to be little more than a museum of 'how it used to be done'. 'Young people want to hear about the exciting possibilities of digital,' they said, 'and we will attract more students that way.' 'Maybe,' I replied 'but you will not be teaching them about the real recording industry.' The fact that the majority of top of the line studios still have truck-loads of analogue equipment that they are not inclined towards parting with, fell on deaf ears. They eventually found another designer, which was perhaps just as well, as they were hoping to tout a designer's name as part of the attraction; something which I would never allow by a studio with whose philosophies I could not agree.

Digital, of course, will be a huge part of the future, but to run 'comparison' recordings from side-by-side analogue and digital control rooms does not provide any comparison at all. Their digital rooms were to be pre-programmed, to show how fast digital equipment could be used in recordings, especially when many changes were needed, but they were not counting the time of preparation in their final analysis. They were also planning to spend equal amounts on the analogue and digital rooms, as an example of the price to performance compromises. This is also not realistic, because with the relatively low cost equipment that they were intending to use, the comparison bore no relationship to the upper levels of the recording industry, where the cost of making digital equipment mimic the quality and speed of use of the finest analogue equipment can be very expensive indeed. In some cases prohibitively so.

There is little doubt that at the lower levels of the price ranges digital can usually out-perform analogue, but the situation rapidly changes when the overall system quality rises. For a *school* to teach that analogue is out of date and in its death throes, merely to attract more clients, is an outrage. Unfortunately, though, the world being what it is, it is a fact that it is easier to teach exciting things on cheap digital equipment than on cheap analogue equipment. However, this could lead to a generation of recordists who have been taught nonsense, simply because it is more profitable for schools to do so. Can we control this? It is an important question.

14.5 The art, and the noise

One thing that most computers have in common is that they tend to produce a considerable amount of heat and noise. The noise is usually from two sources, disc drives and cooling fans. Particularly in many computer-based studios, the computers seem to take pride of place, but their noises have no business in a professional recording control room.

It may seem amazing that it is digital recording that has been selling itself on its dynamic range and low-noise capabilities, yet I have measured background noise levels as high as 48 dBA in computer-based sound control rooms that were charging for their services. Thirty-five dBA is about the reasonably acceptable limit for background noise in a control room for serious purposes. Ideally this would be less, 20–25 dBA, but since the earliest days of recording, mechanical devices, such as tape recorders, have sounded their presence. Realistically though, one needs a good 70 dB between the peak monitoring level and the noise floor, or too many problems will go unnoticed. I realise that there is a great convenience factor in having tape machines, disc drives and other frequently used data storage systems close to hand, but their noise intrusion can only be allowed to go so far before action must be taken to find another point of compromise in terms of operational flexibility. This has led to many control rooms having adjacent and easily accessible machine rooms, with remote controls and metering in the control room itself.

14.5.1 A solution

I often hear people say that they absolutely *must* have the disc drives in a convenient place, which tends to mean not having to move out of their chair. Somehow, though, the really professional studios, who *must* work under pressurised conditions, still manage to site most of their disc drives in a machine room. In small studios, I am constantly being told that there is no way to work with the disc drives out of the room, because they cannot be separated from the computers. Well, this is not a professional response. Whilst it may be true that disc drives can be difficult to separate from the computers, it is usually only the keyboard, mouse and monitor which absolutely need to be close to the operator, and I have rarely encountered any problems in extending the leads to such devices by up to 15 metres. This is usually sufficient to find a suitable place for the computers and their disc drives. It may be a *little* less convenient for the operators if the computers are expelled from the control rooms, but there are many other demands made of control room performance, and everything else is by no means subservient to the ease of operation of the computers.

In short, it is absolutely out of order to pollute the monitoring environment with the noise of disc drives and cooling fans. There *are* solutions to these problems, and if they are not put into effect, then a studio will have to accept that it is *not* capable of working to the standards which can be genuinely called professional.

14.5.2 Similar problems

As we are discussing heat and noise intrusion here, it may be worth mentioning the similar problems with power amplifiers and power supplies. In almost all cases, the power supplies for mixing consoles of any moderate size will be relatively large, noisy and hot. There are rarely *any* problems in acquiring and using extension leads which allow the power supplies to be housed in other rooms, or even in corridors. With the power amplifiers for the monitors, though, such options are rarely available.

For best sonic performance, power amplifiers need to be close to the loudspeakers, and rarely should loudspeaker cables of more than 2 metres be used. Cable inductance *can* impair sonic quality, and *cannot* be reduced by using fatter cables. Although there are in many cases expensive solutions to this problem, it is true to say that below lengths of two metres, almost any cables of adequate size will not impair the sound quality of most loudspeakers likely to be used in studios. To apply the 'two metre rule', though, tends to mean mounting the monitor amplifiers inside the control room, but this is also useful if it is necessary to see any warning lights.

I really do not like using fan-cooled amplifiers in control rooms, because even if the fans are quiet when new, they often soon become undesirably noisy. Convection cooled amplifiers may often be larger, and somewhat more expensive, but at least they are almost silent. It seems absurd to have to point out that we are discussing *sound* control rooms, in which it tends to be a good idea to reduce extraneous noises to a minimum if quality work is to be performed, but such is the case. Noisy amplifiers simply have no place in sound control rooms.

14.6 Conclusions

There is no doubt that computers have made a great impact on the recording industry, throughout almost all of its diversity. At the highest levels, computers are used to great advantage, but they almost always remain subservient to their masters, the recording engineers. At the domestic level, many home studios have only been made viable by computer-based technology and virtual mixing consoles. Discrete, digital consoles have already made their mark, and can provide excellent value for money. They can also be great fun. In the domestic set-ups, time pressures are rarely great, and nobody is going to have an orchestra in their living room, so the limitations on the speed of operation are usually of little concern.

In the middle range of semi-professional and project studios, though, there has been a tendency for the computer systems to dictate the working practices and environments, and their noisy intrusions are often accepted as an inevitability. It is in these circumstances where great care is often needed to keep the computers in their places. If the restrictions are such that the best recording practices for any type of music can *not* be met, then it is only fair to make the musicians aware of the fact. They should not be made to believe that 'this is the way to do things' merely because the studio cannot supply the requirements for recording in the ideal way. If the musicians go along with the recording restrictions, then all is fine, but fragile creativity should not be jeopardised simply to get a recording done at all costs.

Computers have become a great boon to the recording industry in general, but one should never forget that it is the humans who should be quite firmly in control of the how's and why's.

Chapter 15

Considerations on music-only surround: can we control the chaos?

In the cinema world, there tend to be standardised formats for surround which not only cover the mixing processes, but also the conditions for the performances. This has ensured that a respectably uniform appreciation of the wishes of the directors and/or mixers have been presented to the audiences. The subsequent transfer of these soundtracks into the realms of home theatre, though, is something which cannot be expected to be so uniformly achieved, because the variety of playback systems and listening rooms ensures that there will be an element of hit and miss in all this. However, the results, in general, appear to have been satisfactory, so far.

There are no doubt three main reasons why the current situation appears to be working well. Firstly, film production is very expensive, so the mixing of feature films tends to be done by experienced engineers with professional attitudes. Secondly, the novelty of home theatre and the new sensations which it brings to domestic environments are creating a level of enjoyment which people already feel to be satisfactory. However, thirdly, and very importantly, there is a visual domination of the whole event. When watching a film, the listener/viewer rarely worries about a dB here, or half a dB there. In fact, 2 dBs here or 3 dBs there are also unlikely to cause too much distress. However, with the mass sale of music-only surround looming ever nearer, the 'microscopes' of the audiophiles will be poised to ensure that the critical assessment of high-end hi-fi standards will soon be brought to bear on surround. Five- or 5.1-channel surround frees us from many of the limitations of stereo and offers us a whole new range of possibilities, but it does bring along its own problems, one of which is the optimisation of how to reproduce the low frequencies most accurately.

In 5-channel surround, using five full-range loudspeakers, we have five possible low frequency sound sources: left, centre and right front, and two rear loudspeakers. In relatively anechoic listening conditions, as would seem to be optimum for surround, the frequency response at the listening position would largely be that of the axial response of each loudspeaker for any sound source routed to one loudspeaker only. One of the beauties of surround is that it offers a discrete centre-front channel, which not only anchors the sound positionally, but gives us a more predictable reproduction in a variety of listening conditions. This subject is discussed in Chapter 12.

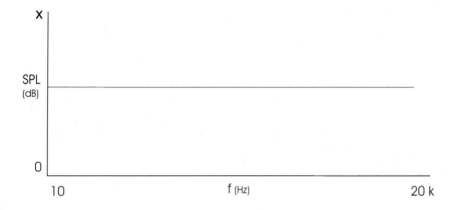

Figure 15.1 Frequency response of a perfect loudspeaker in an anechoic chamber

A mono source drives a room in a much simpler manner than a phantom stereo source because there is much less interaction with the complex modal patterns, but even in an anechoic chamber, or any other free-field listening conditions, a phantom source from two loudspeakers *cannot* give a uniform response except on the centre line which runs between them. A perfect single loudspeaker in a perfectly anechoic chamber would produce a frequency response as shown in Figure 15.1, and would have a polar pattern as shown in Figure 15.2. If we were to replace our single source with a pair of sources, in order to derive a phantom centre image, then we would still achieve the frequency response of Figure 15.1 on the centre line between the sources. However, if we were to move off axis, then the responses typical of those shown in Figure 15.3 would result. This is because the path-length differences to the two loudspeakers would cause comb-filtering, as different frequencies arrived with different phase relationships. This fact is well known, but it is highlighted by the quite startling polar patterns shown in Figure 15.4. This shows the combined polar patterns for two loudspeakers reproducing the same signal, at the same level, as would be the case when reproducing a central phantom image in stereo.

If we were to transfer our loudspeakers into a more typical set of listening conditions, such as would be the case for most of the music buying public, then room reflexions, even on-axis, would interfere in such a way that a response somewhat like that shown in Figure 15.5 would result. Only at low frequencies, where the wavelengths are long, would the interference remain wholly constructive. The result would be a 3 dB rise in low frequencies, relative to the upper frequencies, up to a frequency which would be wholly determined by the *distance between* the loudspeakers. This is due to acoustic mutual coupling between the loudspeaker sources, where the pressure produced by each loudspeaker superimposes itself on the diaphragm of the other loudspeaker. As moving coil cone loudspeakers are constant velocity sources, more work will be done when the added pressure from the other loudspeaker

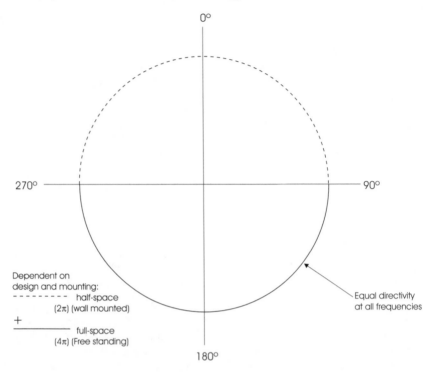

Figure 15.2 Polar response of a perfect monopole loudspeaker in an anechoic chamber

is resisted (in order to maintain constant velocity), and more work done will result in more power being radiated as the cones become more resistively loaded.

15.1 Surround panning

Over the years, we have learned to live with these problems in stereo recording and reproduction, though they are the cause of many sound compatibility

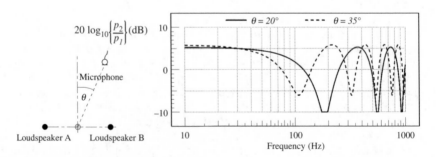

Figure 15.3 Off-axis frequency response of a pair of 'perfect' loudspeakers when producing a central phantom image in an anechoic chamber

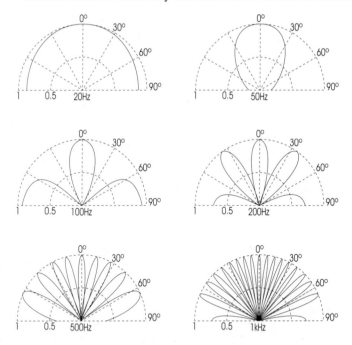

Figure 15.4 Polar response of a pair of 'perfect' loudspeakers when producing a central phantom image

problems from one environment to another. They have also been the source of much frustration and time wasting for those who did not understand what was going on. All too many people have seen stereo as a simple solution to everything (if they could only get it right), whereas in reality, it is actually a

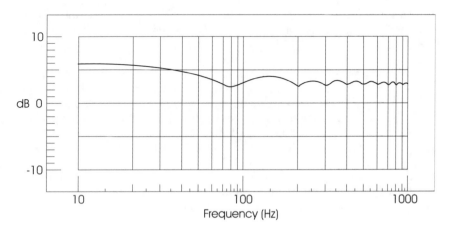

Figure 15.5 Frequency response of a pair of 'perfect' loudspeakers when producing a central phantom image, on-axis, in a typically reflective room. The 0 dB level would be the response for one of the loudspeakers in anechoic conditions

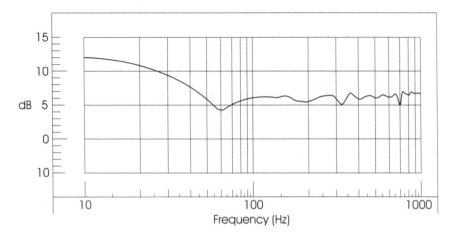

Figure 15.6 Frequency response of four 'perfect' loudspeakers simultaneously reproducing the same programme in a typical room

complex solution to very little. Nevertheless, the complexity of stereo pales when compared to the complexity of the interactions between surround loudspeakers. Stereo supports *one* phantom sound stage from two loudspeakers: surround can support *ten* phantom sound stages between any two of the five loudspeakers, and *twenty-six* when one signal is shared by three, four, or five loudspeakers.

As the same rules of acoustics apply in stereo and in surround, then perhaps we should look at what happens in a situation where a signal is panned into an in-room location, say by being fed to the four corner loudspeakers. The same rules are in force as produced the plot shown in Figure 15.5 for a stereo pair, but here, with four loudspeakers interacting, we would get a response as shown in Figure 15.6. The problem has clearly now become serious, as a 6 dB low frequency boost when panning from a corner position into the room will be unlikely to be subjectively acceptable. What is more, it would not be possible to 'equalise as you pan', because the upper limit of the low frequency boost is solely dependent upon the distance between the loudspeakers, so the end result would change from one home to another. The response at different listening positions would also be a greatly variable factor for a signal panned to all four corner loudspeakers. Figure 15.7 shows the patently low-fi response plots for a listener at two positions in a room other than the exact centre between the loudspeakers. Once again, the problems are significantly worse than in the corresponding situation for a stereo pair of loudspeakers. If such a 'creative' use of surround sound were to be envisaged, then clearly the use of a single sub-woofer would be desirable, where the flat frequency response from a single source could once more be enjoyed.

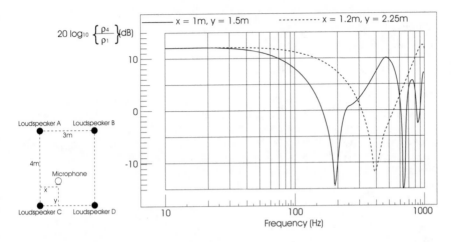

Figure 15.7 Combined frequency response of four omni-directional loudspeakers at two different, non-central positions in an anechoic chamber

15.2 The single sub-woofer option

There is still much discussion about the subjective quality differences between the use of full-range stereo loudspeakers, and the use of 'satellites' and a single sub-woofer. It has certainly been *my* experience that separate, full-range sources sound more natural, but the difference reduces with the frequency of crossover. I prefer to restrict the sub-woofers to frequencies below 80 Hz, though to let them go down to 15 Hz or less. One hundred hertz appears to be about the upper limit before dislocation of the respective sources becomes apparent.

On classical music, though, the situation appears to be somewhat different. When surround is used to maximise the sensation of being in a large hall, the low frequency wash seems to be a rather important aspect of the sensation of realism. It has also been noted that with organ music, recorded in churches, some very odd phase artefacts can be noticed in the low frequencies, which close in if a single sub-woofer is used. I personally believe that a more natural reproduction is achieved by the use of separate low frequency sources for each position of the surround, but this *only* holds true in listening environments with well controlled low frequencies. As low frequency reflexions increase, the situation rapidly changes in favour of the single sub-woofer approach. From looking at Figures 15.6 and 15.7, it is little wonder why this should be the case.

We thus appear to have a situation whereby discrete low frequency sources can scale greater heights of fidelity *if* – and I repeat, if – recordings are made and mixed with due regard to the electro-acoustic constraints of a 5-channel system. If ignorance prevails, then a 5.1-channel system is a safer option. However, it should be noted that due to mutual coupling and multiple-source effects, the optimum electrical balance of a mix, at low frequencies, will differ in the 5- or 5.1-formats. So, which way should we go? Let us consider what is good recording/mixing practice within the 5-channel limits.

15.3 Mixing considerations

The strength of surround comes from it having five (or more) discrete sources of sound. This can be further enhanced by ceiling loudspeakers, but let us restrict ourselves to the current format, for now. The weakness of surround comes in the number of possible phantom stages that can be supported – twenty-six, or so! The question is, given the five discrete sources, do we need to use the phantom sources for anything of great importance, and, if so, can we maximise the potential and limit the damage? So much of the compromising stereo has been as a result of trying to over-exploit what it had to offer. Similar greed in the use of surround sound can carry even greater grief potential. Surely the best use of surround is to provide a 3-channel stereo frontal sound stage with ambience in the rear loudspeakers. Any centrally placed instruments or voices, whether in classical or more modern music, should be fed only to the centre-front channel. We then have two possibilities for the centre-to-left and centre-to right images. We can pan them between left and right, as in conventional stereo, or we can pan them between either side and centre. Psycho-acoustic consideration would tend to suggest that the former would be better for on-the-centre-line listening positions, and the latter for off-the-centre-line positions, perhaps trading stability of the image for some perceived clarity. The subtended angle of the left/right pair could also have an influence on the choice.

When single sub-woofers are used it is often good practice to mount them off-centre, so that low frequency room modes are not symmetrically driven. This practice tends to cause more peaks and dips in the overall response in the room, but of less severity than the fewer superimposed response irregularities caused by symmetrical driving. When this is done, I recommend offsetting the sub-woofer to the right, because when listening to orchestral music, this is where the basses will be coming from. Much has been published, and much talk is bandied about, on the subject of the lack of directional information below 300 Hz, but this early work was done with sine waves. In the case of music, arrival times and the shapes of transient wavefronts carry much directional and timbral information, and I can guarantee that in almost all cases, except in acoustically poor rooms, an orchestral bass section on the right with a sub-woofer on the left will *not* sound as natural as with the sources co-located. Obviously, with a centre-panned bass guitar, the offset to the left or right is equal, so the choice for orchestral music will cause no further compromises. This is just one of many aspects of low frequency reproduction to be taken into account that are so often ignored. And what of the case of discrete rear-positioned instruments with a single front sub-woofer? In such cases, different listening positions can cause unacceptable arrival time differences which can muddy the response. There are thus many points which need to be considered further before launching blindly into audiophile surround mixing.

Certainly in the cinema world, it is relatively unusual for any signal to be fed to the front *and* surround loudspeakers simultaneously, except, perhaps, for an explosion, or something similar, where the full power of the whole system is needed for effect. The cinema people also tend to refrain from putting discrete images in the surround loudspeakers, because they do not want people to be tempted to look away from the screen. It will almost certainly be the exploitation of these 'taboo' practices by the music-only surround mixing personnel which will cause the greatest problems in surround compatibility, and the greatest disappointments in their domestic reproduction.

What the music-only mixers must realise is that acoustic summation of the outputs of surround loudspeakers in non-ideal, domestic circumstances, and the electrical summations which they will need for fold-down into stereo and mono can be wildly different. If they do not fully respect this fact, and I doubt that most of them will, then they will only have their own ignorance to blame when they find that their surround mixes will not travel well to other systems. A careless mix will also be unlikely to fold-down to stereo with the instrumental and tonal balances intact, and what will come out of a mono radio is anybody's guess.

15.4 Centre-front – the new advantage

There is also another point which people will have to get used to when wishing to work to audiophile standards for music-only surround. The single most important loudspeaker is the one in the centre-front position. The nonsense of using a smaller loudspeaker here, and considering it to be a dialogue-only channel, is a spin-off from video productions, but for music-only it is folly in the extreme. This loudspeaker needs to handle, alone, things such as bass guitars, bass drums and vocals, which are currently often split as a phantom image between a left/right pair. If anything, the centre loudspeaker should be *twice* the size of the left or right loudspeakers, not *half* the size. Centre-front is now our priority channel. We are thus faced with a conflict of interests, here, between the needs of high quality audio and some of the established surround practices in use in the multi-media world. For example, a bass guitar reproduced as a phantom central image at 95 dB will have its power divided between the two loudspeakers of the pair, and will also gain a little extra at low frequencies from mutual coupling between the sources. Each loudspeaker of the pair therefore needs only to put out 92 dB at the most, and in reality, perhaps a little less. On the other hand, a single, central loudspeaker will itself have to generate the full 95 dB output. The trend towards using a *smaller* loudspeaker in the centre-front position is therefore at odds with the requirements for the highest quality of audio reproduction. The film industry is also now moving in the direction of taking maximum benefit from the centre-front loudspeaker, so some careful consideration will soon need to be given to the question of the compatibility of multi-format releases where perhaps only one mix is expected to be used for film, home theatre, video or music only releases. The front loudspeakers should, in most cases, be identical.

As we have seen in Chapter 12, when a sound emanates from a stereo, phantom source, its frequency response as perceived by a pair of ears will not be the same as that of a sound reproduced by a single, central loudspeaker. Because of path-length differences to each ear from each of the sources of a phantom image, cancellation will occur in the region around 2 kHz, where the inter-aural distance is about half a wavelength. By using a single loudspeaker centre source, a flat response (at least within the capability of the loud-speaker) can finally be reliably delivered to the listeners. This effect is obvi-ously not detected when measuring in-room responses with a single microphone, and because it is rarely seen in measurement, its presence is often not duly recognised, even by some experienced recording personnel who *should* know.

By positioning the important centre-stage instruments in a discrete central loudspeaker, the stability of the stereo image also becomes much more robust. When a listener moves off the centre-line, left, right *and* centre images will remain firmly locked in place, with none of the tendency for the centre image to collapse into the nearest loudspeaker to the listener. Phantom sound stages between the left-and-centre and right-and-centre will also not be able to col-lapse across the entire sound stage, at least not if side-to-centre panning is used. The prime advantage, though, is that central images are produced by *one loudspeaker only*.

Multiple loudspeakers reproducing the same sound, as is the case with a central image from a stereo pair, or an in-room surround signal, will produce significant interference effects within the room. Except on the central plane between the loudspeakers, all the room boundaries will receive double signals, separated in time (and hence in phase, also) by an amount which relates to the different distances from any point on the boundary to the two sources. The effect of these differences are dependent upon the shapes, sizes and surfaces of the rooms in which the music is heard, and thus the room-to-room inconsistencies of sound quality are much greater for phantom sources than for mono, discrete sources. This all reinforces the need for rather absorbent listening conditions for surround, both for monitoring and for end-users alike.

If special mixing rooms need to be specially constructed for audio-only surround, then so be it. Many stereo sound recording studios have been adapted for surround use, and there has been a tendency to compromise centre-front loudspeakers for the sake of better vision into the studio, but how long the situation can remain is uncertain. Once audio-only surround becomes more commonplace, dedicated surround sound-only mixing rooms will be needed, just as the film industry uses dedicated mix-only rooms. This may lead to a situation where the video people are left out on a limb with their concept of the compromised centre-front channel. In fact, further 'clarifica-tions' may well be needed when it comes not only to the appropriate use of centre loudspeakers and sub-woofers, but also to the choice of rear loud-speakers.

15.5 How to surround?

The rear loudspeakers make their own different demands for the different concepts of reproduction. If the audio-only people *are* going to want to place discrete instruments in the rear channels, then good full-range loudspeakers will be needed here. This is another area which may conflict with the film and video world, though there are good reasons why the audio-only mixes should *not* exploit this option, except, possibly, as in the video/film worlds, when special or occasional effects are required. Ambient surround needs good dispersion of the sound, and may well be best generated from relatively diffuse sources. Distributed mode panel loudspeakers seem to show some potential in this area. 'Hi-fi' sound from the rear, however, tends to demand discrete, localised sources, but under most common circumstances, this is fraught with problems.[1]

As can be seen from the above, brief look at some of the different priorities for music-only surround, there are a number of areas which need multidisciplinary discussions and assessment if we are not soon to find ourselves in a situation where the film, video, music, specialised home theatre, and possibly other interests are not to all go their separate ways, leaving a confusing overall situation for the consumers, and expensive multiple mixes for the producers. The very diversity of use which DVD has to offer will soon force many decisions upon us, and it would be a pity if the whole thing degenerated into the sort of incompatibility nonsenses which led to such a fiasco with the quadrophonic recordings of 20 years ago. We have been warned, so if we fail, there will be no excuse.

15.6 Hi-fi or surround?

There are therefore some difficult dilemmas which now face studio designers when they are asked to design control rooms for audio-only surround formats. The lack of standardisation within the industry on the precise distribution of loudspeakers and frequency bands does nothing to ease the design of rooms which are required to cater for a number of different concepts of what surround should be. When the rooms must also be stereo compatible, the problems increase still further.

There is a very serious probability that many surround-sound control room design concepts do little other than provide a performance stage for overzealous musicians, producers and engineers to amuse themselves. Given the level of ignorance about the pitfalls which await unrestrained music-only surround mixing, any chances for the record-buyers hearing what the recording personnel hoped that they would hear are rather slim. System to system, and room-to-room compatibility problems have always dogged stereo, but these problems compound themselves enormously when the number of loudspeaker channels increases.

One of the greatest problems, though, seems to be the reluctance, on behalf of many people, to recognise that the problems even *exist* to any significant degree. Record companies largely only see the possibilities for re-releases in surround as huge money-earners for little outlay. Equipment manufacturers see surround as a great way to sell more equipment, and many recording staff, although well-intentioned, plunge headlong into something which simply will not do what they are expecting it to do. Audiophile quality surround has not really been a point under serious discussion, and no amount of money which may be spent on home hi-fi surround systems can create realistic effects if the recordings are flawed. It would appear to be the case that hi-fi and surround are unlikely to accompany each other.

15.7 Fundamental requirements for stereo

The greater part of the past 30 years of my life have been spent working as a recording engineer, producer and studio designer. During that time, I have usually been looking for, or seeking to design, control rooms in which I could be sure that what I have been hearing was a reasonably accurate representation of what was being recorded on tape. If there were problems, then I wanted to hear them so that they could be solved before the recordings reached the shops. If the recordings sounded great, because they *were* great, then I wanted to know that, also. Monitoring which lacks adequate resolution of fine details will tend to allow defects to pass to the end-products in the shops, which will be unsatisfying for the discerning customers. Monitors which flatter, suggesting that things are better than they really are, will tend to lead to disappointing results when the recordings are played in rooms other than those in which similar flattery may take place. As an engineer or producer, that would be unsatisfactory for me, as well as for the home listeners. Neither of the above results could be considered to be professional. Professional recording studios are not simply ones that charge money, but are those that have an attitude of professionalism in their operation. Mercenaries get paid for doing a job, but professionals get paid for doing a job to the best standards achievable under circumstances which allow some fundamental standards to be upheld. A brain surgeon cannot do a good, professional job with blunt and dirty instruments. Even with the finest instruments, though, I would be unlikely to allow a *butcher* near my brain. The right skills, *and* the right tools, are needed to do such a job properly. Be it in the realms of brain surgery *or* recording, skill, experience, knowledge and good equipment are all required if professional and consistent results are to be achieved. Putting all of this together, in stereo, has been no easy feat, but putting it all together in surround can be a nightmare. Bearing these facts in mind, studio designers generally try to achieve monitoring conditions which have uniform, or smoothly tapering frequency responses, an absence of early reflexions which can colour and confuse the sound, and low levels of distortion and background noise, which both tend to mask the musical subtleties. Achieving these things in surround can be *very* difficult, even in controlled professional environments, and in practice may not be possible to realise domestically.

Almost all studio designers would agree on the above basic requirements, though many other aspects of high quality monitoring may well be given different weightings in their individual hierarchies of priorities. These hierarchies have led to different concepts of stereo control room design such as 'Live-End-Dead-End', 'Non-Environment', and other approaches which seek to allow a clean 'first pass' as the sound initially reaches the ears, so that uncoloured, uncluttered monitoring of the signal can be achieved. In many cases, the biggest area of contention between designers has been the finding of the optimal balance point between definition, stereo imaging, and providing an ambience which is both comfortable for the musicians, and ostensibly acoustically 'representative' of some perceived average of domestic listening conditions. At the root of all this has been the fact that stereo cannot simultaneously support all the demands that have often been unreasonably made of it.

The main perceptual differences between two of the concepts are that the Non-Environment (NE) rooms are biased towards quality control inasmuch

as they allow the revelation of very fine sonic detail and excellent low frequency uniformity of response. The 'Live-End-Dead-End' (LEDE) rooms
(and some others which seek to achieve a reflexion-free zone around the listening positions, and allow a longer, semi-reverberant decay on the total room
response), can achieve a greater feeling of immersion in the music, which
some musicians feel to be necessary. They can also, in certain cases, generate
stereo images which are more positionally stable over wider listening areas,
though which are not always as pin-point accurate as those perceived in the
NE rooms. The basic concepts of the rooms are shown in Figures 11.3 and
15.8.

A number of properties of the NE and LEDE types of rooms are somewhat
mutually exclusive, and compromises often tend to yield neither one thing nor
another. Nevertheless, with so many different types of people and music
existing side by side, each of the different design philosophies have their
weaknesses, their strengths and their followers. The compromise exists to a
large degree because of the inherent limitations in two-loudspeaker stereo.
The fact is that whatever ambience that may be added to the stereo reproduction by means of room acoustics, either by design or by accident, will not be
the true ambience of the recording space. What is more, any superimposed
ambience will tend to mask detail in the frontal image, and will be different

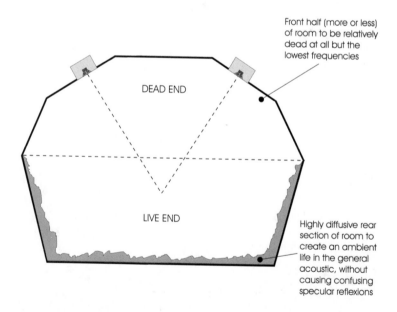

Front half (more or less)
of room to be relatively
dead at all but the
lowest frequencies

DEAD END

LIVE END

Highly diffusive rear
section of room to
create an ambient
life in the general
acoustic, without
causing confusing
specular reflexions

Figure 15.8 The concept of the Live-End-Dead-End stereo control room, in which a reflexion-
free zone is created around the listeners. The rear half of the room is made to be highly acoustically diffusive, usually without specular reflexions, to simulate a sense of ambience within the
room which does not obscure the general clarity of the monitoring. The front half of the room
is acoustically dead, to prevent any room effects from returning from the frontal direction
which could spacially superimpose themselves on the direct monitor sound

from one room to another. What will be heard when people take the music home from the more ambient types of control rooms will thus be no more likely to be in the control of the recording personnel than from mixes done in the NE rooms, which dispense with any idea of 'guessing' the domestic ambience, and which concentrate only on monitoring what is on the recording.

15.8 An answer to our prayers?

The differences in the approaches to stereo control room design can all be solved by the use of 5-channel discrete surround. The detail of the NE rooms can be maintained, whilst the centre-front channel can vastly improve the image stability of the frontal panorama. The rear channels can supply the ambience which the rear halves of the LEDE rooms have sought to provide, but with the option of muting them when detailed quality control monitoring is needed for the front channels. (One cannot easily mute the natural, acoustic ambience of an LEDE control room.) The end result is a room such as is shown in Figure 15.9. For the audiophile, this should be a dream come true.

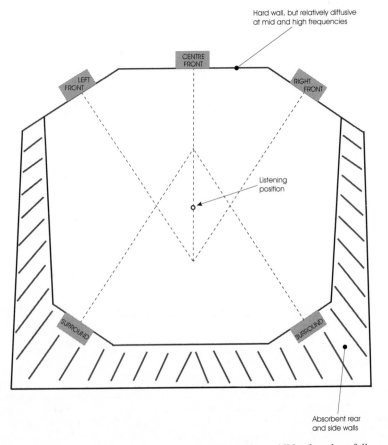

Figure 15.9 Non-Environment room in music-only surround use. All loudspeakers full-range

A 3-channel stereo mix, plus 2-channel surround, can provide a degree of fidelity, image stability and a sense of being there that is entirely unattainable from two-channel stereo. This can be done; it can be achieved in people's homes; and it can also provide a degree of inter-room compatibility that has been so elusive for so long. It is a completely practicable concept.

Unfortunately, though, the ideal mixing styles for best 'hi-fi surround' are unlikely to happen as a mass market option, because 3-channel stereo with ambience is not financially interesting enough for the people who like to make money out of the recording industry. The mad rush is now on to extract every last drop of gimmick potential out of surround, and that is going to mean sacrificing its wonderful possibilities. Basically, the Dolby Digital Surround cinema format is just about optimal, with the only real contentions in its application to audiophile hi-fi being those of whether to use a sub-woofer or three full-range frontal loudspeakers, and, of course, the use of data compression. For true hi-fi, the distribution of the rear channels, either from diffuse or multiple rear loudspeakers, as used in Dolby Digital Surround, is the better way to reproduce ambience. However, the trend in surround *music* studios is for five discrete, identical loudspeakers, which does not conform with the principles of Dolby Digital Surround.

If we take the case of the NE rooms, which lend themselves easily to cinema-type surround use, a layout such as shown in Figure 15.10 differs from that in Figure 15.9 only in that the rear channels in Figure 15.9 are used with discrete, full-range loudspeakers, whereas the cinema arrangement uses more diffuse sources. The Figure 15.10 layout is preferable from the point of view of the ambient nature of the reproduction, but it cannot optimally support a surround mix when there are important solo instruments in the rear channels. For this type of use, the arrangement shown in Figure 15.9 is pre-ferred. Nevertheless, even with the five full-range loudspeakers, we cannot achieve symmetrical front to back monitoring, which would presumably be a prerequisite of 'hi-fi' surround. To some degree, that does not matter, as we do not have symmetrical front to back hearing, but there does exist a problem in terms of room-to-room compatibility.

15.9 Reflexion problems and loudspeaker loading

If the front monitors are flush mounted in a solid wall, then the front wall pro-vides a baffle extension against which the low frequencies can push. This reinforces and flattens the low frequency response. If the rear monitors are similarly mounted (in a solid wall) then the rear of the room would not provide a correct acoustic termination for the *front* monitors, no matter which control room design philosophy was chosen. This was the big mistake made in so many of the quadrophonic rooms of the 1970s, which used two front halves of control rooms, face to face. Control room rear halves are designed to be complementary to the front halves, because without appropriate oppos-ing surfaces, the front halves tend not to work too well. On the other hand, if we suspend the loudspeakers in mid air, then we either create an irregular response due to the omni-directional low frequency (LF) propagation reflect-ing from the wall surfaces behind them, or, if highly absorbent walls are used, as in the NE rooms, we lose a good proportion, up to 6 dB, of the LF output.

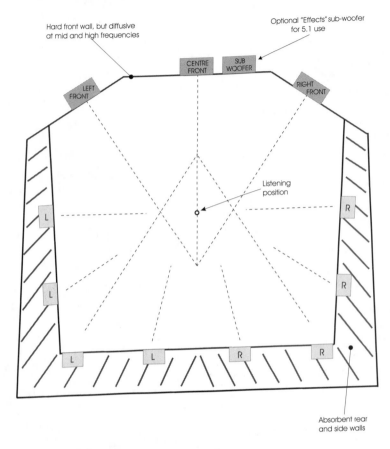

Hard front wall, but diffusive
at mid and high frequencies

Optional "Effects" sub-woofer
for 5.1 use

Listening
position

Absorbent rear
and side walls

Figure 15.10 Non-Environment concept using multiple rear sources. Left and right surround channels split between two groups of four loudspeakers. Alternatively, perhaps diffuse radiators, such as distributed mode panel loudspeakers, could be used. All loudspeakers full-range, except the optional 'effects' sub-woofer

In domestic circumstances, this may mean using amplifiers and loudspeakers with four times the power handling, which would significantly increase the system costs. Marketing people do not want this problem. For full-surround hi-fi, rather than for ambient surround hi-fi, the problems of how to achieve consistent and uniform loudspeaker responses in domestic circumstances are not trivial; in fact, they are almost insurmountable.

In a surround control room, with a hard front wall, any signal which emanates from a rear loudspeaker will strike the front wall and will reflect back into the room, creating a coloured and uneven response. This problem can be ameliorated, to some extent, by making the surface of the front wall very irregular, but it must still act as a baffle extension for the front monitors. Irregular stone seems to be a good choice here, but it can do little for the abrupt termination which it creates for the low frequencies generated from the rear of the room, which *will* be reflected back, creating an uneven response.

In purpose-designed control rooms, this problem can be addressed, to some degree, by the use of pit traps in the front floor, but this would then leave us with a control room which was highly *un*typical of the rooms in which the end-products would be likely to be played, rather than simply being neutral. On the other hand, what is the point of monitoring in typical rooms if they do not lead to hi-fi mixes? The situation seems to be one of 'no win'.

Of course, the intensity of this problem can be moderated by the use of a single, front mounted sub-woofer. Again, in the case of ambient surround, this option is available without too much disturbance to the proceedings, but what with the case of a full-range keyboard instrument in the rear left loudspeaker with its sub 100 Hz output appearing from a front centre-right location? For a listener in the absolute centre of the loudspeakers, equidistant from all of them, there may not be any adverse consequences, but for a listener who moves one metre forward (i.e. one metre nearer to the low frequencies, and one metre further away from the upper frequencies) a delay will be introduced of something in the order of 5 milliseconds between the arrival of the sub bass, and the arrival of the remainder of the sound. Considering the lengths to which high-end hi-fi designers and studio monitor designers go to achieve arrival time accuracy in the thousands of *micro*seconds region, then such *milli*second discrepancies as can be found in only a *slightly* misaligned surround set-up are gross, at least in terms of high fidelity.

15.10 More sources, more problems

So far, we have only considered the reproduction of 'mono' images, from single loudspeakers, but when signals are panned between two or more loud-speakers, in order to achieve mid-speaker or in-room phantom images, the subject of full-surround hi-fi begins to get complicated. When all loudspeak-ers are reproducing individual instruments, or groups of instruments, the signals going to each loudspeaker will be different. They will be uncorrelated, so the interference effects will largely be random. However, when two or more loudspeakers share a common signal, such as when reproducing a phantom image, the interference patterns will have a definite relationship, and will produce distinct patterns of interference. At low frequencies, where the wavelengths are long in comparison to the distances between the loudspeak-ers, the interference will be in-phase, and so will be wholly constructive. At higher frequencies, the summations and cancellations will be frequency dependent, and in terms of level will tend to average to a simple summation of the power. These concepts were discussed in detail in Chapter 12.

What can be achieved in an acoustically controlled room has little in common with what is likely to be achieved in domestic situations in terms of the sound of panned phantom sources, and the more loudspeakers to which a signal is panned, the less likelihood there is that the music will sound similar in different rooms. The mutual coupling effects are controlled by the dis-tances between the loudspeakers themselves, and any reflexions from bound-aries. These conditions, alone, dictate the frequency up to which the mutual coupling boost will be noticeable, and also the degree of boost that will be experienced. Loudspeakers close together will produce a boost up to a higher frequency than loudspeakers further apart, and they will also give a higher level of boost, due to the sounds having less distance to travel, and hence being at a relatively higher level when they arrive at adjacent cones. There are therefore no 'typical' monitoring conditions which can accurately predict the timbre of panned instruments as they will be heard in other, dissimilar listening conditions. As this is a low frequency phenomenon, a quick glance at the equal loudness contours of Figure 5.1 will show that at 1 kHz, a 10 dB change in level will produce a subjective doubling or halving of loudness, but at 100 Hz, a change in level of only about 5 dB, at typical domestic listening levels, will produce the same effect. This could easily yield conditions where the 100 Hz region could be perceived to be twice as loud as the higher frequencies in the musical balance from one set up to another. *This is not hi-fi!*

Judging by what has been said in the press about how people wish to use the possibilities of surround mixing, much of it seems to be driven towards creating sensations, and all too many of the people making these statements clearly do not realise the implications of what they intend to do. There is a big 'buzz' factor for newcomers to surround mixing, and many people tend to get carried away by the imagined possibilities of what they can hear in their particular mixing room. Their enthusiasm and lack of experience can inspire mixes which simply will not travel to other circumstances with any predictability of the balance on reproduction. *This is not professional.*

15.11 Working within limits

Let us look at a list of 'dos' and 'don'ts' in terms of what aspects of mixing techniques can provide relative consistency of reproduction and improved results over stereo, and others which will lead to chaos and much disappointment.

DO ...

1 make full use of the discrete centre channel, putting all central information in the one loudspeaker. The mono, centre source will be much more positionally stable than a central phantom image derived from a left/right pair. The images from a single loudspeaker will also exhibit a smoother frequency response, lower distortion, and greater overall clarity than a phantom image, because, even in a *well* designed room, there are still a few interference problems from dual sources. Furthermore, with single central sources, there are no differing path lengths to the two ears, so the mid-range response dips, inherent in 2-channel stereo (and described in Chapter 12), are completely absent. The room-to-room compatibility of 3-channel stereo mixes is much more robust than that of 2-channel stereo.

2 use the surround loudspeakers for ambience and any special, short-lived effects. When the listening room has a low decay time, the ambience of the recordings can be heard quite realistically. Recordings, well made, either with the natural reverberation from the recording hall or room, or from an artificial reverberation device, will travel well, and give a good sensation of 'being there'. Diffuse sources, or multiple sources such as those used in cinemas, give the most consistent results and widest area of the surround sensation.

3 use a single sub-woofer, below 100 Hz or so, if the listening room or mixing room is not well acoustically controlled. Although, in excellent conditions in larger rooms, full range sources tend to sound more natural, the single sub-woofer option can solve many low frequency irregularity problems in untreated rooms, because they drive the rooms in a more simple manner.

4 try to put bass instruments in, or as near as possible to, single loudspeakers. This will ensure maximum compatibility with whatever end-user format is used, whether using a sub-woofer, sub-woofers, or full range discrete loudspeakers, and it will avoid the horrors of Figure 15.7.

DO NOT ...

1 do any unnecessary panning of images into phantom positions.
2 over-use in-room effects, as these will be unstable and will tend to sound very different in different rooms and on different systems.
3 put important instruments in rear positions. Rear loudspeakers in control rooms will most likely not be representative of domestic rear loudspeakers, where they will tend to face a wall which is reflective at all frequencies.
4 whinge because the realities of surround do not match the marketing propaganda, and disregard these facts.

15.12 Where is it leading?

Manufacturers and record producers alike are frequently tending to believe too much of the hype of surround. This seems to be so similar to the nonsense that we went through with quadrophonics in the 1970s. Nothing has changed about the laws of physics which govern the multi-channel reproduction process, nor has anything changed in human psycho-acoustics. Some of us did learn a lot though – the hard way; but so many of the lessons appear to have been forgotten, or only poorly promulgated. We now have a new generation who are rushing headlong towards the same buffers that quadrophonics hit.

In my opinion, surround is the best thing that ever happened to stereo. It can create wonderful ambience and can help to provide some of the aspects which have been desirable adjuncts to stereo, but which have until now only been achievable by forms of 'trickery' which have often led to undesirable side effects.

Given my seven years' experience of mixing for quadrophonic systems in the 1970s as a recording engineer and producer, I feel that I could still give a very good account of myself in terms of creating interesting surround mixes that were exciting (music allowing) and which could travel well from system to system. However, as a studio designer, I, along with many other designers, face some terrible dilemmas. There seems to be a total lack of industry consensus as to what we should be aiming for, and what consensus there is seems to be demanding of us things which we know cannot work as intended. I can build rooms that can work very well if disciplined mixing is carried out with an understanding of the limitations of surround. I do this sort of work for cinema dubbing theatres, but if I do it for sound only, I am worried that people will criticise my designs precisely because they prevent them from committing too many nonsenses. On the other hand, I can design rooms which give people what they think that they want, and when the results don't travel, I will be accused of building rooms that lie. I get no satisfaction from designing rooms which fail to perform.

So much of what is hyped for surround is gimmickry, and not that which is in the interest of the furtherance of the concepts of hi-fi. Certainly, this is a fact to be faced, because big businesses seek profit, and not altruism, as a primary aim. If giving a new buzz to a huge number of people can be achieved in such a way that there is big money to be made, then such a route will no doubt be followed, and there is nothing wrong with that state of affairs. What *is* wrong, though, is when surround is marketed as an advancement of stereo in virtually all respects. In limited circumstances, it can be an advancement, but those are not the circumstances which tend to be so emphatically marketed, because that is not where most of the money is to be made.

If surround means the use of logic for gain riding, and/or data compression, to squeeze it on to an easily marketed distribution medium, then such systems will always be caught out by skilled ears or certain types of music programme. We could do wonders with DVD by using 21-bit 96 kHz front channels, 16-bit 48 kHz surround channels, and the narrow bandwidth, sub-woofer channel. Such a system could provide *great* 3-channel stereo with ambient surround, without the need for compression or logic. This, in quality terms, could be a huge step forward for domestic hi-fi, and it could offer some excellent new sensations but marketing concerns will probably see to it that such a reasonable approach never gets off the drawing board. The only choice would then seem to be between hi-fi, *or* surround.

Reference

1 Newell, P.R., 'From Mono and Stereo, Through Quadrophony, to Surround', *Proceedings of the Institute of Acoustics, Reproduced Sound 13*, Vol. 19, Part 6, pp. 135–154 (October 1997) (Reproduced in *Audio Media*, European Edition, Issues 85 and 86, December 1997 and January 1998.)

Bibliography

Chase, J., 'Hi-fi or Surround? Part Two', *Audio Media*, European Edition, Issue 92, pp. 122–6 (July 1998)

Hidley, T., 'Full Bandwidth', *Audio Media*, European Edition, Issue 98, pp. 72–6 (January 1999)

Holman, T., 'Audio for Digital Television', *Audio Media*, European Edition, Issue 89, pp. 114–18 (April 1998)

Newell, P. R., 'Hi-fi or Surround? Part One', *Audio Media*, European Edition, Issue 90, p. 188 and pp. 190–2 (May 1998)

Newell, P. R., 'Surround Monitoring', *Audio Media*, US Edition, Issue 9, pp. 54–61 (July/August 1998)

Chapter 16

Horns: their strengths and weaknesses

Perhaps it would be wise, before leaving the subject of loudspeakers and monitoring, to take a short look at some of the properties of horn loudspeakers. Perhaps there is nothing in monitoring so controversial as the use of horns, but in reality, there is a huge amount of misconception and ignorance in many of the arguments employed against their use. Indeed there are some awful horns in use in some monitoring systems, but just as in the case of control rooms, the fact that there are many bad ones out there does not mean that they are all bad. Many people who declare that they do not like horns have merely never heard any good ones, but the fact that they have never heard any does not mean that none exist. In reality, horns do have their uses in highest quality monitoring, but they must be very carefully designed and applied.

16.1 Current uses

In professional sound reinforcement use, where high sound pressure levels (SPLs) are required, there are times when there may be little alternative but to use horns, because a general rule of thumb for almost any loudspeaker systems for high quality music reproduction is that the source of sound should be as small as possible, consistent with wavelength, that is. The higher sensitivity of horn-loaded loudspeaker systems makes physically smaller systems practicable, and can thus help to avoid the terrible timing confusion that would arise from large source areas of less sensitive direct radiators. The directivity control of horns has also been put to good use in such circumstances, by concentrating the sound where it is needed, thus both 'wasting' less power, and avoiding the unnecessary generation of confusing reflexions. Unfortunately, such benefits have often been gained at the expense of sonic purity. This loss is a necessary price to pay in large sound reinforcement applications, and the laws of acoustics deem that it must be so, but the skill of good front-of-house engineers can, with experience, turn adversity into advantage. These people are in the sound *production* business, so they can sculpt and mould the sound to taste. They are not constrained by the need to *reproduce* sound as faithfully as possible.

On the other hand, in cinemas and recording studios, where relatively high levels of quantity *and* quality of sound are required (though not such high

SPLs as in many sound reinforcement applications) the search for high level high fidelity has been pursued with some considerable effort. Perhaps a brief discussion of the current situation may be of interest and guidance also to those in the hi-fi world who are still somewhat perplexed over the continuing horn debate. In fact, the same rules apply, whatever the application. It is merely the compromise points which differ according to the needs of specific situations; the facts set out here are as relevant to studio monitoring and hi-fi, as to any other area of use.

16.2 Physical needs

In systems intended for faithful reproduction of music, horns are only realistically applicable above the 500 Hz region. In many cases, they will not be used below 800 Hz, or even 1200 Hz. Essentially, the physics of low frequency horns makes it virtually impossible, in practice, to get a good transient response from a complete system. Chapter 13 discussed why an accurate transient (time) response requires both an excellent pressure amplitude response (frequency response) *and* an accurate phase response. It thus follows that the arrival times from all the drivers in a system must also be as near to simultaneous as possible, because the time response accuracy is dependent upon the amplitude and phase response accuracy, *and* vice-versa!

Unless a horn is adequately terminated to the listening room, reflexions from the sudden change in acoustic impedance at the mouth will travel back down the horn. These will upset both the amplitude and the phase responses and, in turn, the transient response of the total system output. For a good termination, free of reflexions and response irregularities, the mouth of the horn needs to be as wide as one wavelength of the lowest frequency to be reproduced. At 30 Hz, this would mean having a mouth *10 metres wide*. On large sound reinforcement rigs, sometimes the *combined* mouth area of the stacks of individual bass horns *can* approach this size (they sum very well), and excellent bass can result, but in homes, studios, or even cinemas, such mouth sizes are out of the question. The reflexion problem can easily be demonstrated by tapping the end of a 4 or 5 m length of plastic drain pipe. A 'do yo yo yo yong' will be heard as the sound reflects from each end of the pipe, and produces multiple echoes which can take some time to die away. *Completely* avoiding this effect is the task of a good horn flare termination. When this is not properly achieved, when the compromises have lead to other priorities taking precedence, a faster version of the pipe effect is the cause of much 'horn sound'.

Length, is also a problem. All horns possess what is known as a cut-off frequency, which is the frequency below which the horn will not load the driver resistively, so no 'horn-loading' benefit will result. The cut-off frequency is proportional to the rate of flare of the horn. A low cut-off frequency requires a low rate of flare, so, to flare *slowly* to a 10 m mouth would produce a very long horn. (The underlying reasons for this are detailed later.) However, even if the horn were to be folded, and the mouth size somewhat compromised, a horn for low frequency reproduction would still be some metres long. The sound must travel down the full length of the horn before reaching the listening room, so a delay will be evident between any mid and high frequency drivers and the time taken for the low frequencies to travel down the horn – roughly 3 ms for every metre. The simultaneous arrival of all frequencies required for a good overall transient response, is thus difficult to achieve, at least without resorting to the use of digital delay systems for the higher frequencies. In most situations, such an idea would be a non-starter.

To compound matters still further, just how would one mount the mid and high frequency drivers close enough to the effective source of the low frequencies (the centre of the mouth of the horn) in order to achieve any co-axial, co-located, or even coherent sound source? Large objects placed inside the mouth of a horn will disturb its response, and physical separation of the sources will produce an incoherent sound. It is only around 500 Hz that the physical dimensions of horns become manageable to a degree when they can be incorporated, sanely, into a complete system. A wavelength at 500 Hz is around 65 cm, which is a manageable mouth size, and what is more, the flare rate needed for a 500 Hz cut-off frequency, together with a respectable mouth termination, can allow for a horn length of about 40 or 50 cm. These figures are reasonable from a system engineering point of view. Low frequency 'hi-fi' loudspeaker horns may well sound pleasant to some people, but *true* hi-fidelity (i.e. 'accurate') they are not. The physics of horn loading simply will not allow it. *Nevertheless*, if some people enjoy listening to them, then far be it from me to criticise them. We listen to music for our pleasure, and what pleases is valid. Here, I am only trying to separate the truths from the fiction of what horns can and cannot achieve if the most technically accurate reproduction is required.

Historically, the big research money in horn design has been made by their largest users, the cinema and sound reinforcement industries. The recording studio monitor system manufacturers have largely tended to 'borrow' the sonically better of the ranges of horns produced for other purposes. In this way, though, they have often inherited certain of the more subtle sonic deficiencies left over from the compromises needed for their original applications. In studios, *and* in the hi-fi world, these sonic shortfalls have led to criticisms of 'quackiness' or 'honking', which have become so widespread that many people, even professionals, believe that they are an inherent property of horns, per se. They are *not*! Furthermore, it is widely believed that response aberrations in the cut-off region are unavoidable; this is another 'fact' which is not true. Unfortunately, it has been the case that some loudspeaker manufacturers who do *not* produce horns have seized upon much public ignorance to vilify horns, merely to serve the purposes of promoting their own alternative offerings. Even some very large companies, for whom profit is their sole *raison d'etre,* would readily drop horns if they were seen to be getting too much bad press, and would develop an alternative product merely for the purposes of easier marketing.

16.3 Horn cut-off and the use of exponential flares

Horns are waveguides, which have cross-sectional areas which increase, steadily or otherwise, from a small throat to a larger mouth. An acoustic wave within the horn therefore has to expand as it travels along the horn, from throat to mouth, at a rate dependent upon the local flare rate of the horn. A conical horn is actually a section of sphere. Only at high frequencies and with small size will conical horns be effective. This limits their use as audio horns. At the other extreme, a smooth, parallel, 'plane-wave' tube loads a driver extremely well, but it needs to be of infinite length. If it is cut short, the termination to the air in a room would be very abrupt, so the reflexion problems and frequency response disturbances would certainly not be hi-fi. It would exhibit a version of the aforementioned sound when the end of a plastic drain pipe is struck. However, by employing a horn of an exponential (or thereabouts) flare rate, a relatively uniform, resistively dominated loading can be achieved over a usefully wide frequency range. An exponential flare is one in which the flare rate is constant with distance along the horn. That is, if the distance from point A to point B causes the cross-sectional area to double, then any other points separated by the same distance as A and B will also have a 1:2 relationship in their cross-sectional areas. Whereas in the conical horn the loading *gradually* changes from reactive to resistive as the frequency rises, in an exponential horn, the geometry causes an abrupt change. The transition from reactive to resistive loading takes place simultaneously, throughout the entire length of the horn, at the cut-off frequency. Unlike a spherical expanding wave, in which the loading is dependent upon the frequency *and* the distance from the source, (a situation which applies in conical horns), the impedance loading upon an exponentially expanding wave-area is dependent upon frequency alone: *the cut-off frequency*. Small, short horns, with wide regions of uniform response, can therefore be realised in practice by applying exponential flare rates. Figure 16.1 gives some comparison between conical and exponential performance.

The four points of greatest relevance are:

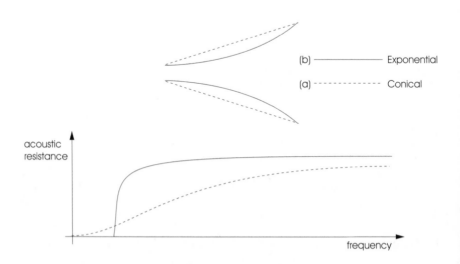

Figure 16.1 Comparison between the normalised throat resistance of an exponential and a conical horn of comparable size (ignoring reflexion from mouth). The acoustic resistance curves are typical of the 'frequency responses' to be expected from such horns

1 The cut-off frequency of the throat depends upon the rate of flare, and the rate of flare alone.
2 The smooth termination from the mouth of the horn to the room is dependent upon the gradual nature of the way that the flare blends into the baffle. It is also dependent upon the *size* of the mouth, the largest dimension of which should not be less than one wavelength of the lowest frequency to be reproduced.
3 Mouth size must be kept within reasonable limits, otherwise it becomes impossible to close-couple the horns to other drivers in a multi-driver system.
4 If horns are more than about 40 cm in overall length, from the diaphragm to the mouth, even low level reflexion problems can become so temporally separated from the initial signal that 'horn-like' artefacts can begin to emerge in the sound character.

The above restraints are quite severe, but unless they are respected the ultimate horn performance is unlikely to be realised.

16.4 Degrees of loading

Professional monitoring horns, such as the Reflexion Arts AX2, shown in Figure 16.2, are designed in accordance with the requirements of horn physics, and are exceptionally vice-free. Figure 16.3 shows the throat impedance of the AX2 through its cut-off region, compared to the response of a very widely used, 'constant directivity' horn, also *intended* for monitoring use, but one which makes compromises for a wider directivity.

All horns are acoustic waveguides, but the term waveguide appears to be being applied more and more to horns of the non-compression type, where a small diaphragm, be it a cone, dome, or ribbon, is placed at the throat of a usually exponential waveguide (horn). (The term waveguide is probably, in most cases, being used both in order to sound technical, and to also fool people who believe that they do not like horns into buying 'waveguide' systems.) By using a combination of a 'waveguide' and a relatively high efficiency drive unit, in-band sensitivities of 100–105 dBs per watt at one metre, or even more, can be achieved without the need for pre-compression. This zero/low compression technique greatly improves the sensitivity, and hence maximum given output SPL from the drivers, without introducing some of the pitfalls of high-compression techniques. Such sensitivities obviously lend themselves well to the use of single ended valve amplifiers, especially for the 1 kHz+ range of bi-amplified systems. However, I am not advocating such use; I am only mentioning it in passing. It may present an interesting possibility for home hi-fi, though.

High frequency loudspeakers generally have a range of sensitivities from about 82 dB @ 1 metre, for 1 watt output, to 112 dB @ 1 metre, for 1 watt input. One watt into the latter sensitivity would be as loud as 1000 watts into the former. A sensitivity of 103 dB @ 1 metre for 1 watt would require 8 watts, on the same scale, so this helps to put the sensitivities into perspective. However, only the use of compression drivers tends to be able to achieve the 106–112 dB sensitivity range, which is why they are so useful for concert use.

Figure 16.2 Axisymmetric horn geometry

16.5 Axial symmetry

There is a large body of opinion, both in studio and domestic circles, who consider *all* mid-range horns, to 'honk', 'bark', 'quack' or sound 'hornlike' in various other ways, yet the Tannoy Dual Concentric loudspeakers were rarely, *if ever*, accused of sounding typically horn-like. Nevertheless, above 1 kHz, or thereabouts, they are horns, through and through. What is more, they were the much maligned *compression* horns, which, by means of a phasing plug, as shown in Figure 16.4, compressed the waves through narrow slots, to further increase the driver sensitivity. Given the state of the art when these devices were first produced, such as the lack of high power amplifiers or high temperature voice coils, a compression driver was used to gain sensitivity, but it was of a relatively low compression ratio. The bass cone was used as a horn flare, and thus, above 1 kHz, the Tannoys were axisymmetric horns of low compression. Interestingly, the 1990s have seen the emergence of numerous loudspeaker designs employing axisymmetric horns, both for studio *and* hi-fi use, in which the horns show remarkable geometrical similarity to the Tannoys of old, not the least of which are the *new* Tannoys. Many of their current models are still using axisymmetric HF horns. When one considers what is being discussed in this chapter, this longevity of the Tannoy design should not be too much of a surprise, because they follow the rules set out in Section 16.3. They do not attempt to use horn loading below 500 Hz, they use axially symmetrical geometry, they have a smooth, gradual mouth termination, (at least when flush mounted) and they are less than 40 cm long. They obey the four cardinal rules for *true* high-fidelity use of horns.

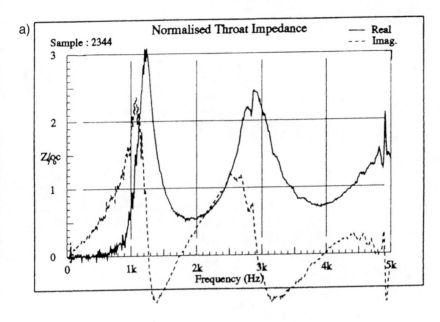

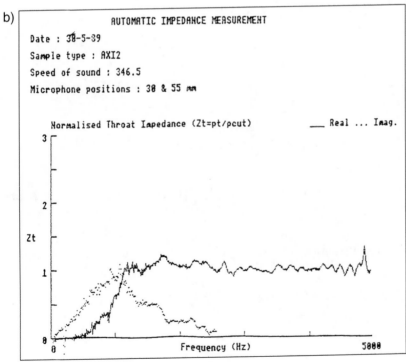

Figure 16.3 (a) Throat impedance plot of a typical constant directivity horn;
(b) throat impedance of AX2 axisymmetric horn (As shown in Figure 16.2)

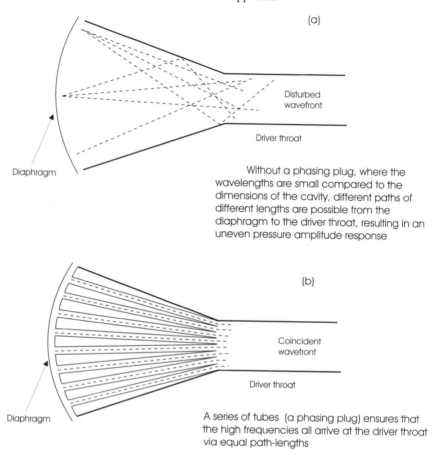

(a)

Disturbed
wavefront

Driver throat

Diaphragm

Without a phasing plug, where the
wavelengths are small compared to the
dimensions of the cavity, different paths of
different lengths are possible from the
diaphragm to the driver throat, resulting in an
uneven pressure amplitude response

(b)

Coincident
wavefront

Driver throat

Diaphragm

A series of tubes (a phasing plug) ensures that
the high frequencies all arrive at the driver throat
via equal path-lengths

Figure 16.4 Throat compression (a) without phasing plug, (b) with phasing plug

The crossover frequency of around 1 kHz ensures that the wavelengths around the crossover point (about 33 cm) do not exceed the mouth diameter of the bass cone. At least this is the case with the famous 15 inch model so widely used by the recording industry. Any disturbances to the smooth flare of a horn in the form of cross-sectional area changes, either in the flare itself or at the termination of the mouth, will cause reflexions which will superimpose themselves on the pressure amplitude response *and* phase response. What is more, any abrupt angles, such as at the junctions of the sides, top and bottom of a rectangular horn, will create off-axis response anomalies in propagation directions opposite to those junctions. This also holds true for any objects such as dividers, or even super-tweeters, which may be placed within the horn flare. Consequently, conventional rectangular horns, especially those with dividers, and also constant-directivity horns (which achieve their aims by means of diffraction from sudden cross-sectional area changes – see again, Figure 16.3) will compromise their ability to produce smooth and even wave amplitude and phase responses. If absolute fidelity is sought, then it would appear that only horns of axial symmetry, with circular mouths, can be realistic contenders.

16.6 Radiation considerations

Figure 16.5 shows the typical radiation pattern from a direct-radiating cone. The diaphragm radiates by a piston-like action, producing sound from the whole of the surface area, almost simultaneously. If the sound is radiating from a relatively large surface area compared to the associated wavelengths, then, depending upon the relative position of the listener and the diaphragm, the same path length differences at higher frequencies will occur as would require a phasing plug to 'straighten them out' in a compression driver. In the case of the direct radiator, instead of the peaks and dips caused by the path length distance interference creating an *overall* irregular response (as in horns), the effect manifests itself as lobing of the polar pattern of the radiation into the listening room. The peaks and dips thus occur not only at different *frequencies*, but also at different *positions* in the room, producing different responses at different listening angles.

In highly absorbent rooms, with few reflexions, the response irregularities can be noticed in the direct sound when moving around the room. In more reverberant or reflective rooms, the effect can be an imbalance in the spectral content of the reflective or reverberant energy, dependent upon the nature of the lobing (peaks or dips in the directional response) where the sound strikes the reflective surfaces before returning into the room. In the case of direct radiators, the general 'fix' for this is to keep the radiating areas small, consistent with the wavelengths of the highest frequency to which any given driver will be used. However, as frequencies rise and the units become smaller, we can run into power handling/headroom problems. Furthermore, because of the need to keep going to smaller drivers as the frequency rises, more drive units, fed from more crossover points, need to be used in many higher output systems. Here we encounter an even bigger problem. Because all of the drivers cannot occupy the same point in space at any one time, they must all radiate from different positions, which disturbs their *overall* polar patterns. To make matters worse, more crossover points usually play havoc with the transient response of the whole system, as crossovers tend not to be minimum phase devices.

For any multi-driver system, there can only be one point in the room which can be equidistant from all the drive units, hence lobing of responses is inherent in the very nature of multi-drive unit systems. No amount of 'Time Aligning', phase correction, or even active signal processing can change this fact. Such processes can only go some way to reducing the significance of the effects. However, a purely co-axial, two-way system, such as the Dual Concentrics, *can* become almost perfect point sources, but once the SPLs rise, the large excursions of the bass cone will inevitably begin to modulate the loading on the high frequency horn termination. The result of this is not the usual frequency dependent or position dependent variations, but *level* dependent response irregularities. Whichever way we turn, at least where *true* hi-fidelity is the aim, we seem to run into one caveat or another when attempting

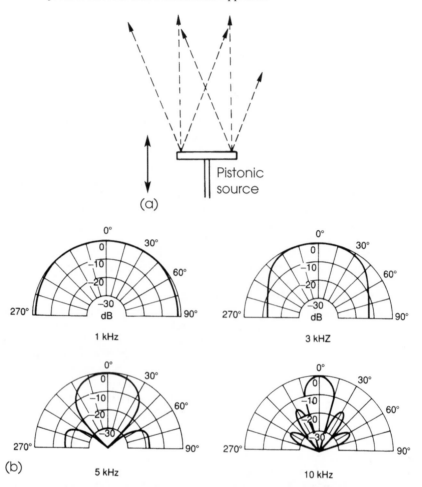

Figure 16.5 (a, b) Typical radiation from a pistonic driver (cone, dome etc.) and the effect on the radiated sound field. (a) Radiating pistons produce particle motions which can interfere with each other, producing interference patterns and subsequent lobing on the polar pattern plots. (b) Typical, lobed radiation pattern of a piston radiator (4½ inch) (after Beranek)

to increase the radiated sound pressure levels beyond a certain point. In *this* aspect of life, designers of domestic hi-fi loudspeakers have a much easier task than the designers of studio monitor systems. Studios tend to require peak levels 10–20 dB higher than domestic users, which means that the systems must be capable of producing 10 to 100 times the acoustic output power (20 dB represents a one hundred times power change). System engineering requirements can be very different indeed when a 100:1 power output differential is required.

Figure 16.6 shows the typical propagation pattern from a typical horn in two planes. What leaves the mouth of the horn, depending upon the precise rate of flare, is a section of a spherical or spheroid expanding wave. Axisymmetric horns produce sound by a means similar to the action of a

(a)

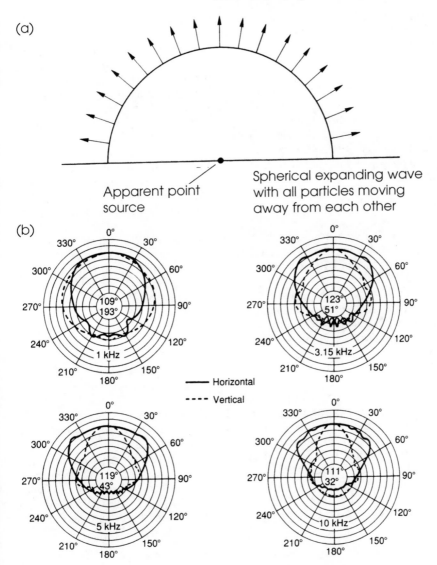

Apparent point
source

Spherical expanding wave
with all particles moving
away from each other

(b)

Figure 16.6 (a) Typical sound radiation from a horn sound. (b) Typical radiation pattern of expanding wave as produced by a well-designed horn. Note the absence of any lobing (Electrovoice)

section of an expanding or contracting balloon, and quite unlike that of the flat(ish) surface of a 'direct' piston radiator. Wherever one moves within the coverage angle of the horn, only one source is apparent, and no path length differences exist between the source area and the listener. The sound propagates radially, as if from a phantom point-source.

By the use of axisymmetric horns for the mid-range and high frequencies, conversion efficiencies can be achieved which allow smaller loudspeakers to produce higher SPLs over a wider frequency range than can be achieved by direct radiating, without any of the drawbacks of conventional rectangular horns. This can greatly facilitate the design of the whole loudspeaker system, and indeed also its interface with the control room. As stated earlier, without a near-perfect termination, the mouth will tend to produce reflexions, which soon begin to cause the overall performance to revert to some of the response irregularities of rectangular horns. It is not possible to over-emphasise the importance of a smooth termination of the flare merging seemlessly into the front baffle. Free-standing loudspeakers are at a bit of a disadvantage compared to flush mounted loudspeakers, because the cabinet is rarely large enough to provide sufficient baffle area for good termination, but this is as true for direct radiators as it is for horns. Small, free-standing loudspeaker cabinets *cannot* produce ripple-free responses with any type of drive unit. For the smoothest response, a wall is an essential baffle extension for *any* type of loudspeaker.

Hopefully, what I have been able to outline here has not been too technically biased. I have tried to make the presentation widely accessible whilst remaining as factually accurate as possible. The point which I am intending to make, most strongly, is that whilst horn loudspeakers are capable of producing the absolutely highest degree of currently achievable sonic fidelity, they can only do so if they are used in ways which respect the laws of acoustics. These laws, and the use of the horns which obey them, are not flexible. The performance of all horns is dictated by the geometrical requirement of the wavelength of the frequencies involved. One cannot scale horns to room sizes or sound pressure levels.

It may well be that people like full-range horns for hi-fi, with all of their inaccuracies, but one will usually find these things to be music related. For example, recordings of music with high levels of ambience, perhaps made via the 'inaccurate' technique of using spaced omni-directional microphones, may sound great on full-range horns. Rapid attack music, with tight bass sounds, is unlikely to do so. However, for high-fidelity in its *literal* sense, horns are capable of revealing low level detail in a way that many other driver systems can have trouble in achieving. Their attack on transient sounds, when used within the rules, can also be of the very highest calibre. A monitor system based on the principles discussed in this chapter is shown in Figure 5.7.

Chapter 17

Foldback

Several of the previous chapters have dealt with a very important aspect of *any* recording studio – its monitoring environment, as this is the means by which all the studio proceedings are ultimately judged. However, we should not forget that other important monitoring environment, the foldback system. When musicians are using foldback, it is their world. The actual sounds of the studios, the sounds of the instruments, or the sounds as heard by the engineers in the control rooms are all only peripheral concepts to a musician wearing headphones. What is of utmost importance at the time of recording is the ambience created for the musicians in the foldback systems.

17.1 A virtual world

In large studios, 'tracking loudspeakers' are sometimes used, as shown in Figure 17.1, but only certain types of room lend themselves to being optimally usable with this form of foldback. The system has always been popular, because many musicians prefer not to use headphones if they can avoid it. Some of this is no doubt because our perception via headphones is different to our perception via our pinnae (outer ears), and musicians tend to like to perform in a familiar word. However, it could well also be that many musicians have shunned the use of headphones not only because they find them to be an unnatural 'world' but also because they have too often had to endure some pretty appalling foldback mixes.

To an alarming number of recording personnel, foldback is something via which a musician keeps in time and in tune to a backing track, and little else. To musicians, the foldback is their creative space. Badly balanced foldback will distort the musicians' perception, and how can anybody be expected to perform with appropriate expression and feeling if they are not hearing anything which is of a nature to encourage such a performance? How can a musician be expected to build up a feel for a track if levels are going up and down, and instruments are cutting in and out as recording engineers adjust *their* balances, while hoping that the musicians can use such run-throughs for their own rehearsals? It is like asking a brain surgeon to perform an operation with the lights sporadically going on and off. Musical performances can be fragile, and this fact should be taken into account before any atrocities are committed via the foldback.

Figure 17.1 Use of tracking loudspeakers. In this situation, a directional microphone is used to record the vocalist, with a loudspeaker providing the foldback. A typical arrangement would be for the loudspeaker to be placed directly behind the microphone, and pointing towards an absorbent surface in order to control the amount of reflexions of the foldback signal returning to the front of the microphone

17.2 Constant voltage distribution

The time-honoured tradition of providing foldback in many large studios has been to use a power amplifier as a constant voltage source, and to use medium impedance headphones to bridge the line. Such a system is depicted in Figure 17.2. The principle of such a system is to take a feed from the auxiliary output(s) of the mixing desk, and to feed this into the inputs of the amplifier(s). The amplifier gain control is usually set such that 0 VU from the desk would produce about 3 dB below clipping at the output terminal of the amplifier. For a typical amplifier producing 100 watts into 8 ohms, the output voltage, at full power, would be given by the formula:

$W = V^2/R$ which can be rearranged as $V^2 = WR$

where:

W = power in watts
V = voltage in volts
R = resistance (or the impedance for a complex load) in ohms

∴ For 100 watts into 8 ohms, the voltage would be:

$V^2 = 100 \times 8$
$V^2 = 800$

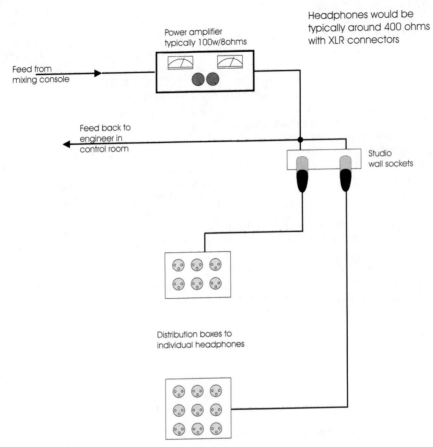

Figure 17.2 Constant voltage foldback system. NB: if individual volume controls are used for each set of headphones, they should be potentiometers of about 600 ohms, linear track, and should each be capable of dissipating at least 5 watts, or they will tend to burn up

$$V = \sqrt{800}$$
$$V = 28.3 \text{ volts}$$

To be 3 dB down on this would be half power, so:

$$V^2 = WR$$
$$V^2 = 50 \times 8$$
$$V^2 = 400$$
$$V = \sqrt{400}$$
$$V = 20 \text{ volts}$$

(Note: 3 dB down is half *power*, 6 dB down is half *voltage*)

The output from a typical power amplifier is referred to as a constant voltage source because when an input voltage causes an output voltage to appear at the loudspeaker terminals, differing load impedances do virtually nothing to cause the voltage to vary. At least, that is, until such an impedance is reached which is so low that the amplifier power output becomes limited by its ability to supply current into the load, when either the voltage will begin to sag, or protection devices cut in.

Going back to our original formula $W = V^2R$ it can be seen that if the voltage remains fixed, then changing R will change W. In fact, reducing R will always increase W. The *power* output is thus dependent upon the load imped-ance. As we add more headphones, in parallel, to a foldback circuit, the impedance of the load to the amplifier will drop, and it will be able to supply proportionately more power. (See Chapter 4 for a discussion on resistance and impedance.) If we choose suitable impedance headphones, then we can ensure that they should not be overloaded if the inputs from the mixing desk outputs are inadvertently turned up excessively, or if a feedback should occur.

A typical studio headphone would be something like the Beyer DT100. Each capsule has a maximum power rating of 2 watts, and they are available in a variety of impedances from 8 ohms to over 1000 ohms. A common value for studio use is 400 ohms. We calculated earlier that an amplifier rated at 100 watts into 8 ohms, could give a maximum output voltage of 28.3 volts. If we substitute 400 ohms into our above formula, we will find that such an ampli-fier could deliver, at maximum:

$$W = \frac{28.3^2}{400}$$
$$W = \frac{800}{400}$$
$$W = 2$$

i.e. 2 watts (into a 400 ohm load).

Consequently, even if driven to maximum output voltage, the amplifier could not deliver enough power to destroy the headphones.

If we wanted to use six headphones, the combination of six 400 ohm loads in parallel would give a total impedance of:

$$\frac{400}{6} = 66.6 \text{ ohms}$$

So, once again, we can use the formula

$$W = V^2/R$$
$$\therefore W = \frac{28.3^2}{66}$$
$$W = \frac{800}{66.6}$$
$$W = 12 \text{ watts}$$

Twelve watts into six headphones is still 2 watts per headphone.

Now we will try the same for twenty headphones:

$$W = \frac{28.3^2}{400 \div 20}$$
$$W = \frac{800}{20}$$
$$W = 40 \text{ watts}$$

which, divided by the twenty headphones is, yet again, 2 watts.

It can thus been seen that if the headphone impedance is correctly chosen, the amplifiers will simply provide more current (and therefore more power) as the number of headphones is increased, until the maximum output rating of the amplifier is reached. However, adding more headphones will not affect the loudness of any of the headphones already connected, nor will there be a risk of blowing up headphones if other parallel devices are disconnected, because for any given input signal, the voltage remains constant, and independent of load impedance.

One precaution which is worth taking with such systems, though, is to feed the foldback amplifiers via un-normalled jacks, so that they must be patched in when needed. All too frequently, people wire them via jacks on the jackfield that are normalled to the auxiliary outputs of the console. During mixdown, when the auxiliaries are being used to drive effects processors, headphones are often inadvertently left connected in the studio. When the effects processors are plugged into the half-normalled auxiliary outputs, the feeds to the amplifiers are not disconnected, so the high levels which may be being sent to the effects units will also be feeding the foldback amplifiers. Although the amplifiers may not be able to overload the headphones with clean power, if the amplifiers do overload, then the distorted signal waveforms, continuously running into the headphones, may destroy the capsules after some time of such abuse. However, by making it necessary to physically patch the amplifier inputs into the auxiliary outputs when foldback is required, the headphones will automatically be disconnected when the auxiliaries are needed for other purposes, and damage to the headphones will be prevented.

With the type of foldback system under discussion, here, all the feeds are sent from the auxiliary outputs of the mixing console. As all the headphones on one circuit share one amplifier, the foldback balance is thus entirely in the hands of the recording engineer. It is usual for the engineer to be provided with a headphone socket in the vicinity of the mixing desk, fed from the same amplifier output, so that he or she will always be able to check, at any time, the exact balance and level that is being fed to the musician(s). In order to optimise this facility, it is the usually preferred method for the engineer to use identical headphones to the musicians.

Of course, not all of the musicians want to hear the same mix, so multiple foldback systems are often provided. Drummers may need to hear lots of bass but little drums in the headphones, whilst bassists may need to hear lots of drums and little bass. No matter how many systems are in operation, there would be a feed from each one to the engineer, so that he or she can carefully set the balance on each channel, in the sure knowledge that the musicians will be hearing the selfsame sound.

17.3 Stereo or mono

Without doubt, a stereo foldback presentation is easier for musicians to work with, compared to an 'above the head' mono jumble. In general, because of their spacial separation, individual instruments can be readily detected in a stereo mix even if their level in the mix is on the low side, or they are in a frequency range which is shared by other instruments which are playing at the same time. Nevertheless, mono foldback is still common, especially when multiple mixes are needed, as there may not be enough auxiliary outputs or foldback channels to send stereo feeds to everybody. Mono mixes usually require much more delicate balancing if the musicians are to hear the detail that they need, and it is often wise not to send unnecessary instruments into the mix.

Some recording engineers, when overdubbing, send a stereo feed to the foldback which is derived from the monitor feed in the studio, plus a little extra boost on what is being recorded. This system can work well, but it must be done in such a way that any muting or soloing on the main monitors does not affect the foldback, or it can be very disturbing and frustrating for the musicians. It may also, at times, be an inappropriate balance, because perception via headphones can be very different to perception via loudspeakers. Once again, though, if the engineer has a headphone feed from the same power amplifier output that feeds the musician's headphones, then the appropriateness of the balance can readily be checked.

17.4 In-studio mixing

An alternative foldback philosophy to the one outlined in Section 17.1 sends line-level sub-group outputs to the rooms where the musicians are recording, and small mixing desks are provided for each musician or group of musicians. Three variants of this concept are shown in Figures 17.3 and 17.4 a) and b). By this means, sets of sub-groups such as drums, guitars, keyboards, vocals, brass and other such sections can be sent to the studio, where the musicians can set their own balance and level, at least to the degree that the subgroups can provide discrete feeds. Whilst this may sound like an ideal system, with much greater flexibility than the constant voltage system, things may not be quite as beneficial as at first sight. The system tends to work very well in project studios where only one or two musicians are recording at any one time. However, with large groups of musicians, there are potential risks of the band not playing together if they are all listening to different balances. To highlight this point, let me put the case of a foldback system where the drums were fed to four foldback channels; say bass drum, snare, tom-toms and overheads. Dependent upon whichever drum was prominent on any particular foldback mix, the individual musician may be led to emphasise different beats. During larger recording sessions, it is usually the producer or musical director who dictates which beats are to dominate, it is not a question of chance, determined only by an individual musician's choice of foldback balance. Setting the foldback sets the mood, and it is a particular skill of experienced engineers and producers to be able to use the foldback balance to drive the musicians to play in a desired way. Remember, if the musicians are

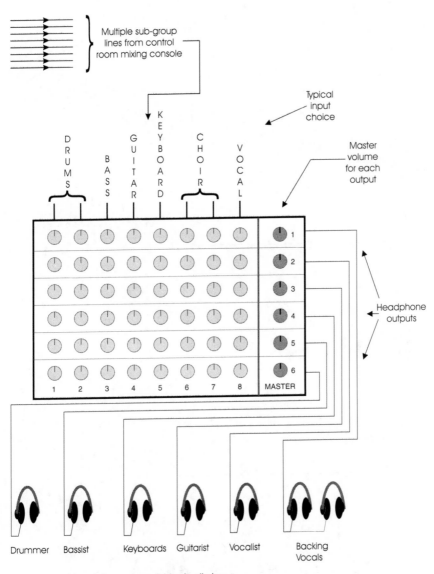

Figure 17.3 In-studio foldback mixing. In this arrangement, single instruments or sub-groups of instruments are fed at line level into the studio. These go to the inputs of a mixer/amplifier matrix which allows each musician, or group of musicians, to set their own foldback mix and overall volume level. The system allows great flexibility in terms of each musician being able to choose an optimum mix, but it needs to be used by musicians who have some skill in mixing, and presumes that they know the score. Its drawbacks are that it splits the concentration of the musicians between mixing and playing, it denies the producer, or engineer, control over the foldback, and it isolates the engineer from knowing what the musicians are hearing

a)

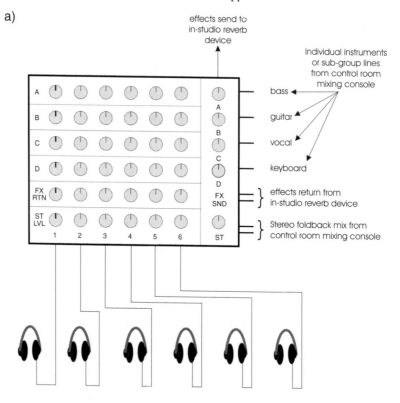

Headphones would be of any impedance between 8 Ω and 1000Ω with stereo jack plugs

b)

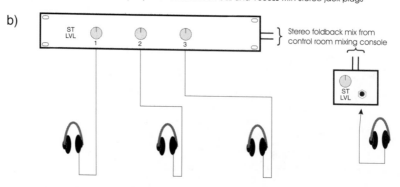

Headphones would be of any impedance between 8 Ω and 1000Ω with stereo jack plugs

Figure 17.4 In-studio foldback mixing (commercially available alternatives). (a) In this arrangement the stereo input level doubles as the output level for each one of the headphones. Effects return level can also be used as a spare stereo input, for a stereo sub-group sent from the control room mixing console. (b) Commercially available single output and multi-output headphones amplifiers. Most of these devices come provided with a mono/stereo switch, and some have an alternative stereo input, which may be used both as effects return or as a second foldback mix input

all going their own ways with the foldback mix, then who is in control of the proceedings? Probably nobody! Quite literally, the musicians may not all be playing to the same beat.

It should also be noted that the musicians do not all necessarily know how to make an appropriate foldback mix, especially of a song with which they are not familiar. Foldback balances are really jobs for the engineers and producers – with frequent reference to the musicians to confirm their comfort with the balances. A problem with individual mixes is that without the ability for the recording staff to monitor the headphones in the control room, a foldback mix can be living a life of its own, which may or may not be relevant to the producer's wishes.

Many project studios have begun with a small, individual foldback mixing system on grounds of cost, and have augmented the number of mixers as the studios have grown. Many studio staff are totally unaware of the constant voltage system because they have never received any professional training. In truth, both systems have their strengths and weaknesses, and when studios develop, there is no reason why they should not offer both options of local and remote mixing. I have known many experienced engineers who have been appalled when they have gone into a studio and heard the mixes that some musicians were actually trying to work with. Conversely, I have known many musicians who have been appalled with what they have been sent from a control room. Getting the foldback balance(s) right is a fundamental necessity for any good recording.

17.5 Types of headphones

Where possible, groups of musicians on a common foldback feed should all be using identical headphones. If mixtures of different impedances are used, or mixtures of headphones with different sensitivities, the loudness of all the sets will not be equal. This is why it is important for the engineer who is monitoring the headphones in the control room to be on the same circuit, and to use identical headphones. However, there is a variation in the preferences of musicians towards open or closed headphones. Many prefer open headphones, because they feel less cut off from their environment, but there can be situations when their use is inappropriate.

Drummers, or other musicians playing loud instruments, may prefer closed headphones. Open headphones provide virtually no isolation from outside noise, so a drummer playing with a kit in a live room, and producing a local 120 dB SPL, will need a potentially ear-damaging level of foldback if using open headphones. On the other hand, headphones with good ear seals may provide 30 dB of isolation, and so from a resulting ambient 90 dB, the drummer could be fed with a mix of instruments which could be heard above the drum level *without* risk to the ears.

One must be careful about the use of excessive level in headphones. The level required in the ear canals for headphones to sound equally as loud as loudspeaker listening can be 6 dB higher. This is probably due to the loss of the contribution which the tactile sense and bone conduction provide as augmentation to the ear-canal-only sound, which means that the ear drum will receive four times the acoustic power from headphones than when listening equally loud to loudspeakers. This extra power can cause unexpected ear damage at levels which do not seem to be subjectively too loud. It should therefore be apparent that if a drummer wears open headphones, the levels necessary to overcome the acoustic 120 dB of the drums by an in-ear subjective 125 or 126 dB, would perhaps, in reality, be subjecting the ears to SPLs of around 132 dB. This is way over any level which could be considered to be safe. For musicians, whose ears are their tools, the results could be disastrous. To put things into perspective, 130 dB represents over 1000 times the acoustic power to which industrial workers would be allowed to be exposed, and whose ears are *not* so necessary for their work.

The above description shows clearly why closed headphones should be used to keep sound out, but vocalists often should wear closed headphones to keep the sound *in*. The problems can occur when overdubbing vocals. Percussion sounds in the backing track can produce the ubiquitous 'tizzy tizzy tizz' that one now hears everywhere from personal stereo systems. If this should leak into the vocal microphone, as often it does, then subsequent processing of the vocal, such as with delays, reverberation or other effects, will also process the over-spill. Especially if the vocal is compressed during the mix, the over-spill will be brought up in level, and the problem becomes worse. Even if a particular vocalist does not like wearing closed headphones, it still may be better to work with one ear-piece partly off the ear, to allow some natural sound in, rather than to run the risk of an undesirable over-spill entering the main vocal microphone. 'Click' tracks entering vocal microphones, via headphones, are also a bane.

17.6 Connectors

When the constant voltage type of foldback system is used, it is common to provide distribution boxes for the headphones, say with six or eight outlets, at the end of a cable of around 4 or 5 metres which plugs into a wall outlet. Outlets can be provided for each foldback channel, either as mono feeds, or stereo pairs, via a simple arrangement of wall sockets, as shown in Figure 17.5. If suitable cable is used, the boxes can be daisy-chained when more outputs from any given feed are needed. Heavy duty microphone cable will usually suffice for such purposes, unless the number of headphones being driven from any single cable would be likely to draw more current than the cable could safely handle. If a tracking loudspeaker were to be used, then it would be better to plug it directly into the wall sockets, which should be wired to the power amplifier outputs with good loudspeaker cable.

The use of 'stereo jacks' on such circuits is not recommended. XLR type connectors tend to be a better choice. The problem with stereo jacks is that when they are inserted into the sockets, they can sometimes short the jack connections together, and a mono jack mistakenly inserted would certainly short circuit the output to one amplifier. Jack plugs are better reserved for use with the individual mixer type of foldback systems. In any case, these systems usually benefit from the use of headphones of much lower impedance that the 400 ohms which is normal for the constant voltage distribution systems. The use of different connectors therefore prevents the erroneous connection of headphones of the wrong impedance into either of the systems.

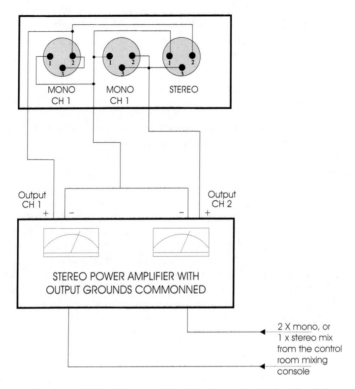

NB: Not all stereo amplifiers can tolerate their output grounds being connected together, but most can. Manuals should be checked before such connections are made

Figure 17.5 Wall socket arrangement for a constant voltage foldback system. In the above arrangement, the headphones are wired with the *–ve* of each capsule connected to *pin 1* of the XLR; the *+ve* of the left hand capsule to *pin 2*, and the *+ve* of the right hand capsule to *pin 3*. When plugged into the mono outlets (or into distribution boxes plugged into the mono outlets) the two capsules of each headset will be connected together, in parallel, from whichever mono channel of the amplifier they are connected. In the stereo outlet, *pins 2* and *3* are connected to different amplifier channels, so the signal in the headphones will appear in stereo.

This is especially important in situations where both types of system are being used in one studio.

As a further precaution against inadvertent shorting of a power amplifier output, it may be feasible to wire a 15 ohm, 50 watt resistor in series with the feed on the headphone outlet wall panel. With only small numbers of headphones in use, this would cause only a minimal voltage drop. However, with 20 headphones in parallel, yielding 20 ohms in total, the 15 ohm resistor would rob almost half of the power, whilst at the same time restricting the amplifier to only half of its output into 20 ohms. It should also be remembered that under such circumstances, the resistor, on music signal, could get very hot as it dissipates around 20 watts of heat. Its siting should therefore be carefully considered. If such a device were to be used for protection of the headphones, then a separate output, perhaps with a 'Speakon' connector, could be wired directly to the amplifier for use when a tracking loudspeaker was needed. The pros and cons of these systems need to be individually evaluated for each set of circumstances.

17.7 Overview

Whatever foldback system may be in use, the importance of providing the musicians with good foldback cannot be over-stressed. Musicians are not instrument-playing robots; they tend to be highly emotional human beings, which is probably why most of them *are* musicians. Creativity can be nebulous, with moods forming and dissolving on some very subtle bases. It should surely come as no surprise that a musician given a superb, spacious, inspiring foldback mix will probably perform more creatively than a similar musician receiving an ill-balanced mono mix via a pair of headphones of indifferent quality.

Tuning of vocals can sometimes be affected by foldback, with the relative loudness of a vocal, in a mix, being able to drive a vocalist sharp or flat. The feel of a song can also be varied, dependent upon which instruments are emphasised. A prominent snare drum, for instance, will tend to draw the musicians to follow that beat, whereas the emphasis of a hi-hat on an off-beat, with the snare subdued, could pull the track in a different direction. I recall one instance when I was producing an album which involved some very famous guitarists of the highest calibre. I was seeking a guitar solo which rose above an extended coda to a song. I knew what I wanted, but such things can be difficult to explain precisely – and especially so in this case as I was looking for a characteristic feel from the guitarists. Several attempts failed to obtain the desired effect, which was finally achieved by giving the guitarist a run-through of a complete demonstration recording, then feeding him only selected tracks of the master. He was fed the master backing track less the bass, drums and instruments which were driving the rhythm. He eventually played over the backing vocal oooh's and aaaah's, a listing rhythm guitar, and very little else. The result was a classic track, where the guitar floated powerfully along, soaring and diving, but cut free from the rigidity of the powerful rhythm. The direction of the guitar recording was entirely steered by the choice of foldback instruments and their relative balance. This could only be achieved by the producer and engineer dictating the foldback mix. This sort of situation can rarely be satisfactorily concluded in cases where the musicians are doing their own foldback mixes.

Remember, so many project studio owners and operators are much more under the influence of equipment manufacturers and dealers than under the influence of top flight studios and their working practices. The option to mix in the studio is readily supplied by ready made boxes, which the manufacturers and dealers want to sell. These can be bought, plugged in with suitable leads, and 'off you go'. On the other hand, the constant voltage/power amplifier option is more labour intensive to install, and may not be as easy to assemble with ready-made items. However, for multiple headphone use, this option will certainly be cheaper for the studio, but it is not necessarily in the interests of the equipment dealers to suggest its use.

Of course, when recordings are being made by relatively inexperienced engineers, the reality is that they may not be too skilful in achieving sensitive foldback mixes, and when working alone, without a producer or an assistant, the delegation of the responsibility of foldback mixing to the musicians may be a great relief. What is more, in such circumstances, the musicians may well do a better job of it, so we cannot be too dictational here about which system is best. Nonetheless, at least if the people are aware that the various possibilities exist, then they have some options when selecting what to do when the time comes for expansion. Of course, some mixing desks do not have sufficient auxiliary outputs to make multiple stereo foldback mixes, and this also must be taken into account when selecting the most appropriate system for any given set of circumstances.

Foldback should never be seen as something of a secondary event in the process of recording; it is crucial. A parallel exists in live sound for concerts. In many countries, I have witnessed the struggle for the prestige of being the front of house (FOH) mixing engineer. True, as an ego trip, FOH engineers are more in the public view, but I have also witnessed the job of stage monitor mixer being relegated, almost literally, to anybody who did not have another task to perform. In truth, the monitor mixing engineer is more likely to have an effect of the quality of the performance than the FOH engineer, for it is the monitor engineer who sets the conditions under which the band must perform. No band is going to play well with poor monitor mixes, and no front of house engineer is going to make a poor performance sound like a good one. In many instances, the calibre of the monitor engineer is more critical to the overall event than that of the FOH engineer. Stage monitors *are* foldback, and when considering foldback in terms of stage monitoring, perhaps the relevance of the foldback mix is easier to understand. In fact, when major touring artistes are travelling the world, their choice of monitor engineer is usually a matter of considerable importance on their lists of priorities. In many cases, the monitor engineer is almost considered to be like a member of the band. The relationships can be very close, because the artistes understand that the monitor mix can be like a lifeline that can either feed or starve a performance. Such is the power of foldback.

Glossary

A/B ing
The switching between different sources, nominally 'A' and 'B', to make comparisons between the sounds. In years past, the line output from a mixing console was usually marked by a switch position 'A', and the return from the tape recorder by position 'B'. Switching back and forth was the usual means of assessing the quality of the recording.

ADAT
Alesis **D**igital **A**udio **T**ape. A format for digital 8-track tape recorders, initially launched by Alesis in the early 1990s. The system uses modified half-inch S-VHS video transports, running at four times standard speed.

A to D
Analogue to **D**igital converter. A device using a highly stable internal clock which samples the audio voltage waveform at a rate higher than at least twice the highest frequency of interest, and puts out a digital binary signal which represents the voltage level of each sample. See also **D to A**. For example, to sample a maximum frequency of 8 kHz, the sampling rate would need to be in excess of sixteen thousand samples per second.

Bridging
A term used to describe the connection of high impedance inputs across a much lower impedance output. This technique is designed not to load the output by any degree that would change its voltage when loaded or unloaded. This was to allow signals to be sent to multiple destinations, of varying number (up to a specific minimum impedance in each case), without disturbing the signal voltage as loads were connected or disconnected.

DAT (or R-DAT)
Digital **A**udio **T**ape. A digital tape format developed by Sony as a domestic replacement for the Compact Cassette. Originally, there were R-DAT and S-DAT formats, the R and S standing for **R**otating and **S**tationary heads, respectively. The S-DAT was ultimately a non-starter (not to be confused with DCC (digital Compact Cassette) which uses data compression) so the R-DAT differentiation has become somewhat redundant, and DAT is now the common abbreviation. The R-DAT and S-DAT concepts were both based on 16-bit

linear, 44.1/48 kHz sampling. It was never conceived as a professional format, but was adopted as such by the recording industry.

DEADSHEETS
Heavily damped flexible sheets, usually of between 8 and 15 Kg/m^2, used in panel damping and membrane absorbers. Bituminous roofing felt was one of the earliest materials to be used as such, but specifically manufactured dead-sheets are now available.

D to A
Digital to **A**nalogue converters receive digitally coded signals, representing voltage waveforms, and by means of clocking and filtering, produce an analogue output voltage which should be as close a representation as possible of the waveform represented by the digits. See also **A to D**.

EDAC
Multi-pin connector widely used in audio installations. They also mate with 'Varelco' connectors, which use the same layout.

Headroom
The margin, usually needed in analogue audio circuitry and tape machines, which represents the region between the nominal operating level and the onset of unacceptable levels of distortion. The headroom tends to be used to allow harmless but necessary transient signal peaks to pass through a system, relatively cleanly, at levels higher than the generally level for more steady-state signals.

Impedance
Any mixture of resistance and reactance which generally impedes the flow of a signal. 'Resistance' acts equally on AC and DC currents, independent of frequency. 'Reactance' is the resistance to the flow of AC, as created by inductance and capacitance, and is highly frequency dependent. Impedance includes properties of both resistance and reactance, in varying amounts.

Intensity
'Sound intensity' is a very specific term, and represents the flow of acoustic energy. It is measured in units of watts per square metre, and should not be confused with either sound power, sound pressure level (SPL) or loudness.

MDM
Modular **D**igital **M**ultitrack machines. These are units of usually 8 tracks which are compact, usually inexpensive, and entirely self-contained with analogue and digital inputs and outputs. ADATS and the DA38, DA88, DA98 units are typical examples.

Minimum phase and non-minimum phase
A signal is said to be minimum-phase when its correction in the amplitude, or 'real' domain also tends towards correcting the response in the phase, or 'imaginary' domain. Minimum-phase responses are typical of gradual roll-offs in filter circuits, mechanical roll-offs, or loading effects on transducers,

due to the radiation into different spaces, for example. The essential factor is that no appreciable delay is involved between the generation of the signal and the effect of whatever is influencing it. If there is no appreciable delay, then there can be no appreciable phaseshifts, hence minimum-phase.

'Non-minimum phase' responses are those where amplitude correction, alone, cannot correct any phase disturbances. The far-field response of a loud-speaker in a reflective room is an example of a non-minimum phase effect. Here, there is a delay between the signal generation by the loudspeaker and the superimposition of the boundary reflexions on the overall response. Reflexion arrival times create phase irregularities which are frequency and distance (time) dependent, so no simple manipulation of the amplitude response of the source can adequately compensate for the complex distur-bances.

Another example of a non-minimum phase effect is in the combination of the various outputs of crossovers. In any filter circuits, either mechanical or elec-trical, there are inherent 'group delays' for any signal passing through them. The amount of group delay increases as the filter frequency lowers, and as its order (6, 12, 18, 24 etc. dB/octave) increases. A crossover will thus have dif-ferent group delays associated with each section, and when the outputs are re-combined, they will **not** produce an exact replica of the input signal. For this reason, conventional equalisation cannot be used to correct for response errors at crossover points. Amplitude correction will lead to further phase, and hence, time response, errors.

RCA/phono plugs (jacks)
RCA jacks and phono plugs are the American and British terms, respectively, for one and the same connector. RCA plugs and phono jacks are confused terms, but are in widespread international use. A similar situation applies with the terms 'patchbays' and 'jackfields' – US and UK usage, respectively.

SPL
Sound Pressure Level. SPL doubles or halves with every 6 dB change, unlike the sound power, which doubles and halves with 3 dB changes. In the acoustic and electrical domains, sound power equates to electrical power, and SPL to voltage. Subjective loudness tends to double or halve with 10 dB changes: 10 dB higher being twice as loud, and 10 dB lower being half as loud. See also **Intensity**. Ten decibels relates to a ten times power change.

Index

www.focalpress.com

Visit our web site for:
- the latest information on new and forthcoming Focal Press titles
- special offers
- our e-mail news service

Join our Focal Press Bookbuyers Club

As a member, you will enjoy the following benefits:
- special discounts on new and best-selling titles
- advance information on forthcoming Focal Press books
- a quarterly newsletter highlighting special offers
- a 30-day guarantee on purchased titles

Membership is free. To join, supply your name, company, address, telephone/fax numbers and e-mail address to:
Elaine Hill
E-mail: elaine.hill@repp.co.uk
Fax: +44(0) 1865 314423
Address: Focal Press, Linacre House, Jordan Hill, Oxford OX2 8DP

Catalogue

For information on all Focal Press titles, we will be happy to send you a free copy of the Focal Press Catalogue.

Tel: 01865 314693
E-mail: carol.burgess@repp.co.uk

Potential authors

If you have an idea for a book, please get in touch:

Europe
Beth Howard, Editorial Assistant
E-mail: beth.howard@repp.co.uk
Tel: +44 (0) 1865 314365
Fax: +44 (0) 1865 314572

USA
Marie Lee, Publisher
E-mail: marie.lee@bhusa.com
Tel: 781 904 2500
Fax: 781 904 2620